Barbara McClintock

Geneticist

Women in Science

Rachel Carson
Author/Ecologist

Dian Fossey
Primatologist

Jane Goodall
Primatologist/Naturalist

Maria Goeppert Mayer
Physicist

Barbara McClintock
Geneticist

Maria Mitchell
Astronomer

WOMEN in SCIENCE

Barbara McClintock

Geneticist

J. Heather Cullen

CHELSEA HOUSE
PUBLISHERS
A Haights Cross Communications Company
Philadelphia

CHELSEA HOUSE PUBLISHERS

VP, New Product Development Sally Cheney
Director of Production Kim Shinners
Creative Manager Takeshi Takahashi
Manufacturing Manager Diann Grasse

Staff for BARBARA McCLINTOCK

Editor Patrick M. N. Stone
Production Editor Jaimie Winkler
Photo Editor Sarah Bloom
Series & Cover Designer Keith Trego
Layout 21st Century Publishing and Communications, Inc.

A Haights Cross Communications ⟋ Company

http://www.chelseahouse.com

First Printing

1 3 5 7 9 8 6 4 2

Library of Congress Cataloging-in-Publication Data

Cullen, J. Heather.
    Barbara McClintock / J. Heather Cullen.
        p. cm.—(Women in science)
Summary: Presents the life and career of the geneticist who in 1983
was awarded the Nobel Prize for her study of maize cells. Includes
bibliographical references and index.
    ISBN 0-7910-7248-7 HC   0-7910-7522-2 PB
    1. McClintock, Barbara, 1902– —Juvenile literature. 2. Women
geneticists—United States—Biography—Juvenile literature.
[1. McClintock, Barbara, 1902- 2. Geneticists. 3. Scientists. 4. Nobel
Prizes—Biography. 5. Women—Biography.] I. Title. II. Series: Women
in science (Chelsea House Publishers)
QH437.5.M38 C855 2002
576.5'092—dc21
                                                              2002015549

Table of Contents

Introduction
Jill Sideman, Ph.D.   6

1. A Satisfactory and Interesting Life   12

2. Self-Sufficient from the Start: 1902–1917   18

3. Early Work at Cornell: 1918–1927   30

4. Choosing a Career: 1927–1941   50

5. Free to Do Research: 1941–1967   72

6. Recognition at Last: 1967–1983   94

7. Barbara McClintock's Legacy   104

    Chronology   110

    Bibliography   113

    Further Reading   114

    Index   116

Introduction

Jill Sideman, Ph.D.
President, Association for Women in Science

I am honored to introduce WOMEN IN SCIENCE, a continuing series of books about great women who pursued their interests in various scientific fields, often in the face of barriers erected by the societies in which they lived, and who have won the highest accolades for their achievements. I myself have been a scientist for well over 40 years and am at present the president of the Association for Women in Science, a national organization formed over 30 years ago to support women in choosing and advancing in scientific careers. I am actively engaged in environmental science as a vice-president of a very large engineering firm that has offices all around the world. I work with many different types of scientists and engineers from all sorts of countries and cultures. I have been able to observe myself the difficulties that many girls and women face in becoming active scientists, and how they overcome those difficulties. The women scientists who are the subject of this series undoubtedly experienced both the great excitement of scientific discovery and the often blatant discrimination and discouragement offered by society in general and during their elementary, high school, and college education in particular. Many of these women grew up in the United States during the twentieth century, receiving their scientific education in American schools and colleges, and practicing their science in American universities. It is interesting to think about their lives and successes in science in the context of the general societal view of women as scientists that prevailed during their lifetimes. What barriers did they face? What factors in their lives most influenced their interest in science, the development of their analytical skills, and their determination to carry on with their scientific careers? Who were their role models and encouraged them to pursue science?

Let's start by looking briefly at the history of women as scientists in the United States. Until the end of the 1800s, not just in the United States but in European cultures as well, girls and women were expected to be interested in and especially inclined toward science. Women wrote popular science books and scientific textbooks and presented science using female characters. They attended scientific meetings and published in scientific journals.

In the early part of the twentieth century, though, the relationship of women to science in the United States began to change. The scientist was seen as cerebral, impersonal, and even competitive, and the ideal woman diverged from this image; she was expected to be docile, domestic, delicate, and unobtrusive, to focus on the home and not engage in science as a profession.

From 1940 into the 1960s, driven by World War II and the Cold War, the need for people with scientific training was high and the official U.S. view called for women to pursue science and engineering. But women's role in science was envisioned not as primary researcher, but as technical assistant, laboratory worker, or schoolteacher, and the public thought of women in the sciences as unattractive, unmarried, and thus unfulfilled. This is the prevailing public image of women in science even today.

Numerous studies have shown that for most of the twentieth century, throughout the United States, girls have been actively discouraged from taking science and mathematics courses throughout their schooling. Imagine the great mathematical physicist and 1963 Nobel laureate Maria Goeppert Mayer being told by her high school teachers that "girls don't need math or physics," or Barbara McClintock, the winner of the 1983 Nobel Prize in Medicine or Physiology who wrote on the fundamental laws of gene and chromosome behavior, hearing comments that "girls are not suited to science"! Yet statements like these were common and are made even today.

I personally have experienced discouragement of this kind, as have many of my female scientist friends.

I grew up in a small rural town in southern Tennessee and was in elementary and high school between 1944 and 1956. I vividly remember the day the principal of the high school came to talk to my eighth-grade class about the experience of high school and the subjects we would be taking. He said, "Now, you girls, you don't need to take algebra or geometry, since all the math you'll need to know will be how to balance a checkbook." I was stunned! When I told my mother, my role model and principal encourager, she was outraged. We decided right then that I would take four years of mathematics in high school, and it became my favorite subject—especially algebra and geometry.

I've mentioned my mother as my role model. She was born in 1911 in the same small Southern town and has lived there her entire life. She was always an unusual personality. A classic tomboy, she roamed the woods throughout the county, conducting her own observational wildlife studies and adopting orphaned birds, squirrels, and possums. In high school she took as many science classes as she could. She attended the University of Tennessee in Knoxville for two years, the only woman studying electrical engineering. Forced by financial problems to drop out, she returned home, married, and reared five children, of whom I'm the oldest. She remained fascinated by science, especially biology. When I was in the fourth grade, she brought an entire pig's heart to our school to demonstrate how the heart is constructed to make blood circulate; one of my classmates fainted, and even the teacher turned pale.

In later years, she adapted an electronic device for sensing the moisture on plant leaves—the Electronic Leaf, invented by my father for use in wholesale commercial nurseries—to a smaller scale and sold it all over the world as part of a home nursery system. One of the proudest days of her life was when I received my Ph.D. in physical and inorganic chemistry, specializing in quantum mechanics—there's the love of mathematics again! She encouraged and pushed me all the way through my education and scientific career. I imagine that she was just like the father of Maria Mitchell, one of the outstanding woman scientists profiled in the first season of this series. Mitchell (1818–1889) learned astronomy from her father, surveying the skies with him from the roof of their Nantucket house. She discovered a comet in 1847, for which discovery she received a medal from the King of Denmark. She went on to become the first director of Vassar College Observatory in 1865 and in this position created the earliest opportunities for women to study astronomy at a level that prepared them for professional careers. She was inspired by her father's love of the stars.

I remember hearing Jane Goodall speak in person when I was in graduate school in the early 1960s. At that time she had just returned to the United States from the research compound she established in Tanzania, where she was studying the social dynamics of chimpanzee populations. Here was a young woman, only a few years older than I, who was dramatically changing the way in which people thought about primate behavior. She was still in graduate school then—she completed her Ph.D. in 1965. Her descriptions of her research findings started me on a lifetime avocation for ethology—the study of human, animal, and even insect populations and their behaviors. She remains a role model for me today.

And I must just mention Rachel Carson, a biologist whose book *Silent Spring* first brought issues of environmental pollution to the attention of the majority of Americans. Her work fueled the passage of the National Environmental Policy Act in 1969; this was the first U.S. law aimed at restoring and protecting the environment. Rachel Carson helped create the entire field of environmental studies that has been the focus of my scientific career since the early 1970s.

Women remain a minority in scientific and technological fields in the United States today, especially in the "hard science" fields of physics and engineering, of whose populations women represent only 12%. This became an increasing concern during the last decade of the 20th century as industries, government, and academia began to realize that the United States was falling behind in developing sufficient scientific and technical talent to meet the demand. In 1999–2000, I served on the National Commission on the Advancement of Women and Minorities in Science, Engineering, and Technology (CAWMSET); this commission was established through a 1998 congressional bill sponsored by Constance Morella, a congresswoman from Maryland. CAWMSET's purpose was to analyze the reasons why women and minorities continue to be underrepresented in science, engineering, and technology and to recommend ways to increase their participation in these fields. One of the CAWMSET findings was that girls and young women seem to lose interest in science at two particular points in their pre-college education: in middle school and in the last years of high school—points that may be especially relevant to readers of this series.

An important CAWMSET recommendation was the establishment of a national body to undertake and oversee the implementation of all CAWMSET recommendations, including those that are aimed at encouraging girls and young women to enter and stay in scientific disciplines. That national body has been established with money from eight federal agencies and both industry and academic institutions; it is named BEST (Building Engineering and Science Talent). BEST sponsored a Blue-Ribbon Panel of experts in education and science to focus on the science and technology experiences of young women and minorities in elementary, middle, and high school; the panel developed specific planned actions to help girls and young women become and remain interested in science and technology. This plan of action was presented to Congress in September of 2002. All of us women scientists fervently hope that BEST's plans will be implemented successfully.

I want to impress on all the readers of this series, too, that it is never too late to engage in science. One of my professional friends, an industrial hygienist who specializes in safety and health issues in the scientific and engineering workplace, recently told me about her grandmother. This remarkable woman, who had always wanted to study biology, finally received her bachelor's degree in that discipline several years ago—at the age of 94.

The scientists profiled in WOMEN IN SCIENCE are fascinating women who throughout their careers made real differences in scientific knowledge and the world we all live in. I hope that readers will find them as interesting and inspiring as I do.

1

A Satisfactory and Interesting Life

I've had such a good time, I can't imagine having a better one.
. . . I've had a very, very satisfactory and interesting life.
—Barbara McClintock

Barbara McClintock was popular with both the men and women in her life. Her mind was vibrant and alive. She could and did play football with the boys in her youth. She was blessed with physical beauty. She lived a long time—she died when she was 90 years old. In those years she traveled across the country, to Europe, and to Mexico; she filled her life with remarkable achievements. She is considered a founder of modern genetics, one of the best minds that field has ever seen—and, of all history's talented geneticists, one of the field's few true geniuses. When she was 81 years old, she won the Nobel Prize in Physiology or Medicine for her discovery of "mobile genetic elements"—a recognition long overdue.

A highly independent person throughout her life, Barbara McClintock did not at first intend to study genetics. Once the field had captured her imagination, though, she pursued it with a clarity of purpose that would lead her to excellence and a Nobel Prize. After some 70 years spent contentedly in the maize fields, she is recognized as one of the most important researchers in the history of genetics.

This is the shed at Cornell, now called the McClintock Shed, in which much of her early research was conducted. She would later remember these days at Cornell—the "golden age" of maize genetics—as some of her happiest. Maize (corn) plants are studied there to this day.

Born to a doctor and a poet, she was raised by free-thinking parents in an atmosphere that allowed her mind to grow. From the earliest age, she was self-reliant. Her mother later remembered that even as a baby Barbara could be left alone: "My mother used to put a pillow on the floor and give me one toy and just leave me there. She said I didn't cry, didn't call for anything." (Keller, 20) They encouraged her to develop other unusual personality traits, too—she often played football with the boys in the happy youth she spent in Brooklyn, New York. Possessed of a vibrant intellect as well as an active spirit, she matured into an attractive, popular young woman. Early on, while still in the college she had chosen for herself, she found a subject that would interest her all of her life—genetics, a science that at the time was still relatively unexplored. For the next 70 years, she did her own research and as a result greatly influenced the development of the field. In those years her work took her across the United States, to Europe, and to Mexico. When she was 81 years old, she won the Nobel Prize in Physiology or Medicine for her discovery of "mobile genetic elements," which she referred to as "jumping genes." Today she is remembered as a genius, one of the three greatest thinkers in the history of genetics.

This she achieved because, most of all, she loved what she did. Some of her contemporaries found it almost scary how much enthusiasm she could bring to what she was doing, to the problem she was set upon solving, to the experiment she was about to carry out. Late in her life she told a story to show how she could get carried away:

> I remember when I was, I think, a junior in college, I was taking geology, and I just loved geology. Well, everybody had to take the final; there were no exemptions. I couldn't wait to take it. I loved the subject so much, that I knew they wouldn't ask me anything I couldn't answer. I just *knew* the course; I knew more than the course. So I couldn't wait to get into the final exam. They gave out these blue books, to write the exam in, and on the front page you put your own

name. Well, I couldn't be bothered with putting my name down, I wanted to see those questions. I started writing right away—I was delighted, I just enjoyed it immensely. Everything was fine, but when I got to write my name down, I couldn't remember it. I couldn't remember to save me, and I waited there. I was much too embarrassed to ask anyone what my name was, because I knew they would think I was a screwball. I got more and more nervous, until finally (it took about twenty minutes) my name came to me. (Keller, 36)

It was about 60 years later, on October 10, 1983, that the Nobel Assembly of Sweden's Karolinska Institute announced the award of the Nobel Prize in Physiology or Medicine to Barbara McClintock for her discovery of "mobile genetic elements." McClintock was the third woman to be the sole winner of a Nobel Prize since the awards were first given, in 1901. The first woman to win in the category of Physiology or Medicine, she became part of a very elite group of women: it had only two other members, Marie Curie, who had won in 1911, and Dorothy Crowfoot Hodgkin, who had won in 1964. Both Curie and Hodgkin had won the prize in Chemistry.

To receive the prize, which included both a medal and a monetary award of 1.5 million Swedish kronor, or about $190,000 in U.S. currency—an enormous amount of money for a woman who had once worried over where she might find employment—McClintock traveled to Sweden. There she and the recipients of prizes in other categories were honored at a ceremony held on December 8, 1983. It is said that when King Carl Gustaf of Sweden presented the Nobel Prize to Dr. McClintock the applause from the audience became so loud and went on for so long that the floor shook.

The press release from the Karolinska Institute in Sweden

McClintock receives the Nobel Prize in Physiology or Medicine from King Carl Gustaf of Sweden in 1983. She won the award for the discovery of "mobile genetic elements," only one of the many contributions that have earned her a place as a founder of modern genetics.

explained the importance of McClintock's work as follows:

> [The discovery of mobile genetic elements] was made at a time when the genetic code and the structure of the DNA double helix were not yet known. It is only during the last ten years that the biological and medical significance of mobile genetic elements has become apparent. This type of element has now been found in microorganisms, insects, animals and man, and has been demonstrated to have important functions. (Nobel)

Certainly a part of this remarkable life was due to her especially strong will and the self-confidence that she had enjoyed since childhood. But a good part of it must also be credited to her loving and free-thinking parents. They allowed their daughter to be as she truly was inside and helped her to become what she *wanted* to become.

2

Self-Sufficient from the Start: 1902–1917

THE MCCLINTOCK FAMILY'S EARLY DAYS

Sara Handy McClintock, Barbara's mother, was born in Hyannis, on Cape Cod, Massachusetts, on January 22, 1875. She was the only daughter of an old and well-respected family of New England. Both of Sara's parents traced their ancestors back to the first families who had come to America on the *Mayflower* in 1620. The families of both parents also included members of the Daughters of the American Revolution, an organization that prided itself on the date of establishment of a family in the United States. Even though Sara's father, Benjamin Handy, was a Congregationalist minister and a stern and righteous man, several other members of the family were more adventurous and free-wheeling. Sara's grandfather, Hatsel Handy, had run away to a life at sea at the early age of 12. Captain of his own ship by the age of 19, he had always been considered a fun-loving man with a quick sense of humor. One of Hatsel's other

President Theodore Roosevelt gives a speech in Connecticut in 1902, the same year Barbara McClintock was born. At the time, Victorian or 19th-century moral codes were still the norm; McClintock would feel the prejudice against women in science throughout her early career. McClintock's parents, however, were free thinkers, and they soon had their daughter behaving in unconventional ways that would become lifelong habits.

children had run off to the California Gold Rush in 1849.

On the death of her mother when Sara was less than a year old, the infant Sara was taken to California to live with an aunt and uncle. Later Sara returned from California to live with her stern, widowed father. She grew into an intelligent and highly attractive young girl. She was an accomplished musician, a poet who later published a book of her own poetry, and a painter. She was also strong-willed, a bit adventurous, and willing to back up what she felt with action. In 1898, she disregarded her father's wishes to marry Thomas Henry McClintock, a handsome young man in his last year at Boston University Medical School.

Thomas Henry McClintock's family did not qualify to join either the Society of Mayflower Descendants or the Daughters of the American Revolution. His parents had immigrated to the United States from the somewhere in the British Isles, probably Ireland. Thomas had been born in Natick, Massachusetts in 1876. Sara used money that her mother had left to her to pay Thomas's bills at school, and the newlyweds moved to Maine to set up their first home. They moved often, first from Maine to New Hampshire and then to Hartford, Connecticut. During these first years of their marriage, they had four children: Marjorie (October of 1898), Mignon (November 13, 1900), Barbara (June 16, 1902), and Tom (December 3, 1903).

According to McClintock family legend, Sara and Thomas McClintock had hoped that their third child would be a boy. They'd planned to name him Benjamin, after his maternal grandfather, Benjamin Handy. The girl that had arrived instead they'd first called Eleanor, and that was the name on her birth certificate. Her parents soon dropped the name Eleanor, though, and began to call the child Barbara. The reason for the change remained something of a mystery to Barbara herself:

> I showed some kinds of qualities (that I do not know about, nor did I ask my mother what they were) that made them believe that the name Eleanor, which they considered to be a very feminine, very delicate name, was not the name that I should have. And so they changed it to Barbara, which they thought was a much stronger name. (Comfort, 19)

The McClintocks changed their son's name to suit his personality, too: he'd been baptized with the name Malcolm

Rider McClintock, but before long he came to be known simply as Tom. Barbara would not change her name officially until 1943, and even then it would be only because her father felt that she would otherwise have a great deal of trouble in securing a passport or proving who she really was during the security-conscious years of World War II.

The McClintocks did not feel constrained by the name on the birth certificate—Sara herself had been christened as Grace. They really were "free-thinkers," a term coming into vogue at that time to indicate those who did not abide by the strict social codes of 19th-century American society. Throughout her early life, McClintock was shown by the example of her parents that to think for oneself was acceptable—and desirable. Sara McClintock published a slim volume of poetry in 1935, the epigraph to which encapsulates the attitude of her poems as well as her thought: "Don't it beat all how people act if you don't think their way." (Keller, 18)

McClintock learned to read at an early age, and reading became a great pleasure for her. She grew to be very comfortable being alone, just thinking:

> I do not know what I would be doing when I was sitting alone, but I know that it disturbed my mother, and she would sometimes ask me to do something else, because she did not know what was going on in my mind while I was sitting there alone. (Comfort, 21)

Surely one of the reasons why the McClintocks moved so frequently during the first years of their marriage was the difficulty Dr. McClintock was having in establishing a successful medical practice. He was not the traditional general-practice or "family" physician, but rather a homeopathic doctor—who based a diagnosis on not only the patient's physical symptoms but also her physical, mental, and emotional state. He approached cures differently, too: believing that an illness was often best left to run its course, he relied on medicines less than regular doctors would. Thus he might not give a patient medication to bring down a fever, but rather leave the fever to rage.

With a quickly growing family and little money coming into the house from Dr. McClintock's practice, Sara began to give piano lessons to neighborhood children to make ends meet. The stress began to show on Sara and in the relationship between her and Barbara. Barbara recalled, "I sensed my mother's dissatisfaction with me as a person, because I was a girl or otherwise, I really don't know, and I began to not wish to be too close to her." And this must have been evident to Sara: "I remember we had long windows in our house, and curtains in front of the windows, and I was told by my mother that I would run behind the curtains and say to my mother, 'Don't touch me, don't touch me.'" (Comfort, 20)

In fact McClintock came to feel like a being apart from the rest of the family, like a visitor. She did not feel picked on or discriminated against, though: "I didn't belong to that family, but I'm glad I was in it. I was an odd member." (Comfort, 21) McClintock was a self-sufficient unit. She considered family life nice but not crucial to her happiness; *that* she could find within herself. Family photographs show smiling, happy children together. It was just that Barbara McClintock felt no emotional need for the others.

At about the age of three, McClintock moved to her father's sister's house in Campello, Massachusetts. Her aunt was married to a wholesale fish dealer there. She lived with the couple off and on for several years before she entered school. It made things easier on Sara, and McClintock remembers the time with joy. She used to ride with her uncle in his buggy to buy fish at the market and then around town to sell the fish door to door. When her uncle bought a car to replace the horse and buggy, McClintock became fascinated with the constant mechanic work necessary to keep the vehicle working. One of McClintock's two major biographers, Evelyn Fox Keller, one of the world's foremost experts on the intersection of gender and

Brooklyn Borough Hall in the center of Brooklyn, New York. The McClintocks moved to Brooklyn from Hartford Connecticut when their daughter was six years old. The family settled in Flatbush, where the young Barbara attended elementary school.

science, cites a revealing episode: At the age of five, McClintock asked her father for a set of mechanics tools; he bought her a children's set, and she was terribly disappointed. "I didn't think they were adequate," she later recalled. "Thought I didn't want to tell him that, they were not the tools I wanted. I wanted *real* tools, not tools for children." (Keller, 22)

THE MCCLINTOCKS IN BROOKLYN

When McClintock was six, she moved back home and her family moved yet again, from Hartford to Brooklyn, New York. Brooklyn, located at the western end of Long Island, is one of the five boroughs of New York City—Manhattan, Queens, Bronx, and Staten Island are the others. In 1908, Brooklyn was already a city of over a million people. Large numbers of both

African-American and Puerto Rican families were moving to Brooklyn because of the jobs available there. Large apartment houses were being constructed, businesses flourished, and the street life was vibrant and alive. Flatbush, the section of Brooklyn in which the McClintocks lived, was filled with new and large single-family homes. Ebbets Field, the home of the Brooklyn Dodgers, would open there in 1913. The African-American baseball team the Brooklyn Royal Giants played at Washington Park. It was a happy time for the entire family, both because Barbara was back and because Dr. McClintock's medical practice was growing. The economic pressures that Sara McClintock had felt so keenly were eased.

In Flatbush, Barbara was enrolled in elementary school.

PROGRESSIVE EDUCATION

The Progressive Education Association was formed in 1919, after two decades of gathering strength in the United States. In line with Thomas McClintock's views, the Progressive Movement was founded on the belief that children should be creative, independent thinkers and be encouraged to express their feelings. This was a radical belief at this time in American history, when structured curriculum focused on the basic skills ("the 3 Rs"—reading, writing, and "rithmetic") was the norm. The Progressive Movement asserted that learning was a gradual process, with each learning experience building on the previous experience. Many supporters of progressivism believed that schools were too authoritarian and that the set standards of school curriculum should be eliminated in favor of teaching what students desired to learn. The Progressive Movement peaked during the Great Depression but fell out of favor by the 1950s, when critics began to claim that progressive education increased juvenile delinquency. (Shugurensky)

From all accounts, her time in grade school was a joy. Her parents regarded school as only a small part of growing up; they believed children should not have to attend if they did not want to. Dr. McClintock once went to Barbara's school to make sure that the teachers there knew that he would not permit his children to do homework. He told them he felt six hours of schooling a day to be more than enough. When Barbara became interested in ice skating, her parents bought her the very best skates they could find. On clear winter days, with their blessing, Barbara would skip school and go skating in Prospect Park. She would stop only when school let out and she could join the other kids for games of football or tag. A "tomboy" by nature, she excelled in sports and played all kinds of games with the neighborhood boys, despite the fact that she was always small. (She remained petite even as an adult, measuring just 5'1".) Her parents were not at all dismayed by her interest in boys' sports; as always, they supported her in whatever she wanted to do, even if the other parents on the block were shocked or upset. They wouldn't let anyone else interfere with any of their children. One time, McClintock remembered,

> We had a team on our block that would play other blocks. And I remember one time when we were to play another block, so I went along, of course expecting to play. When we got there, the boys decided that, being a girl, I wasn't to play. It just happened that the other team was minus a player, and they asked me would I substitute. Well we beat our team thoroughly, so all the way home they were calling me a traitor. Well, of course, it was their fault. (Keller, 27)

Another time, Barbara, dressed in long, puffy pants known as "bloomers," was playing with the boys in a vacant lot. A neighborhood mother saw her and called her over, telling her she meant to teach Barbara how to act like "a proper little girl." Barbara never set foot inside the house; after looking at the woman she simply turned around, went home, and told her mother what had happened. Sara McClintock immediately went to the telephone, called the neighbor, and told the neighbor to mind her own business and never do that sort of thing again.

AMELIA JENKS BLOOMER

Amelia Bloomer, a temperance reformer and advocate of women's rights, became famous in 1851 for the "Turkish pantaloons," called "bloomers," that she'd designed the year before with Elizabeth Cady Stanton. Bloomer wore them with a skirt reaching below the knees. She was not the first to wear clothing of this kind, but her journal, *The Lily*, advocated the bloomers' use and called attention to her. Until 1859, she wore bloomers when she lectured, and she always drew crowds.

She was born in Homer, Cortland County, New York. She married a newspaper editor, Dexter C. Bloomer, of Seneca Falls, New York, in 1840. *The Lily*, which she began as a temperance paper in 1849, contained news of other reforms as well. She also became deputy postmaster of the town in 1849. She said that she wanted to give "a practical demonstration of woman's right to fill any place for which she had capacity."

Bloomer and her family moved to Ohio in 1854 and then settled in Council Bluffs, Iowa. As a suffragist, she continued writing and lecturing for women's rights.

But the reaction to the fashion trend that had been named for Amelia Jenks Bloomer was difficult to defend against: bloomers, which were seen as immodest, stirred the public to anger. One contemporary commentator wrote that trousers on women was "only one manifestation of that wild spirit of socialism and agrarian radicalism which at present is so rife in our land." Barbara McClintock loved to wear bloomers as a girl and would in fact continue to wear pants throughout her life.

ERASMUS HIGH

When Barbara joined her older sisters at the local high school, the school was called Erasmus High School. It had been founded in 1786 as the Erasmus Hall Academy by a group of men that included Alexander Hamilton, John Jay, and Aaron Burr—all key figures in the early history of the United States. Erasmus later became a public school, the second-oldest in the United States. It changed its name in 1896 but is still located at 911 Flatbush Avenue. Today it has been divided into two schools—Erasmus Hall Campus: High School for Science and Mathematics and Erasmus High School for Humanities & the Performing Arts.

Barbara loved Erasmus High, she loved mathematics, and she loved science: "I loved information, I loved to know things. I would solve some of the problems in ways that weren't the answers the instructor expected." She would then solve the problem again, trying to come up with the expected answer. To her, problem solving itself "was just joy." (Comfort, 22)

The McClintock girls had a number of notable classmates at Erasmus High. Norma Talmadge went on to become one of the greatest stars of silent films. She began her work in films at the Vitagraph Studios in Brooklyn in 1910 after class at Erasmus High. Another classmate was Anita Stewart, also a famous actress who got her start at the Vitagraph Studios. Also among the students at Erasmus High around this time was Moe Horowitz, who, along with his brothers Curly and Shemp, created the "Three Stooges" comedy team. He and his brothers made hundreds of films through the 1940s. Some of the famous modern graduates of Erasmus High include Barbra Streisand, Neil Diamond, and former world chess champion Bobby Fischer.

All three McClintock girls did very well in school. The point could be argued whether that was in spite of or *because* of their free-thinking parents and their haphazard attendance. When Marjorie graduated in January of 1916, Vassar College, a prestigious women's college in Poughkeepsie, New York, offered her a scholarship. She did not take it. Sara McClintock did not think that a college education was a very good thing for her daughters. Apparently the family had a female relative who was a college professor and Sara thought she was an old spinster and not very happy—proving her point that proper

LUCY BURNS

Lucy Burns, one of Erasmus High's teachers, went on to play an important role in American women's history. A native of Brooklyn, Burns was born on July 28, 1879. Her parents educated all their children well, regardless of whether they were boys or girls, and they supported Lucy when she went to Vassar College. She started graduate work in linguistics at Yale but went to work for a time at Erasmus before becoming a student at Oxford University in England. It was in England that she became dedicated to fighting, along with other women, to secure the right to vote. She then began to devote her full attention to the cause. She first worked closely with suffragist leaders Emmeline and Christabel Pankhurst. Their Pankhursts' Women's Social and Political Union later gave her a special medal for bravery she'd exhibited—arrested for protesting, she had taken part in hunger strikes in prison to bring attention to the cause. After her return to the United States in 1912, Lucy Burns met Alice Paul. Together they launched a fight to have an amendment added to the U.S. Constitution that would guarantee to women the right to vote. As a leader of the Congressional Union for Woman Suffrage and the National Woman's Party, Burns helped to organize political campaigns, edited a national newspaper entitled *The Suffragist*, and again went to jail. Thanks to the efforts of people like her, in 1920 the 19th Amendment was finally passed.

women just did not go to college. This may have been the prevailing idea when Sara was of college age, but by the 1920s women were demanding and getting a more equal place in society. In 1919, the state of New York offered four-year college scholarships of $100 per year to 205 graduates of high schools in Brooklyn. This number included 71 girls. There were 77 similar scholarships offered to graduates in the borough of Queens, 37 of which went to girls. Her mother's thinking, of course, did not change Barbara's mind.

Barbara graduated from Erasmus High School in 1918, at only 16 years of age. She knew she wanted to go to college, but to her it mattered little where she went. Dr. McClintock had been called to serve as a surgeon in the army in Europe, and his departure meant there was little money coming in. Certainly there was not enough for them to be able to send Barbara to college. Barbara conquered her disappointment and found a job working in an employment agency. She spent her free time reading at the Brooklyn Public Library.

3

Early Work at Cornell: 1918–1927

In the summer of 1918, two events occurred that would allow McClintock to pursue her dream of enrolling in college. First, her father returned home from the war overseas and helped to convince her mother that Barbara's commitment to education was real and they should not stand in her way. Second, she learned that the tuition for the College of Agriculture at Cornell University was free for New York state residents. She applied to Cornell and was accepted. Having overcome the major obstacles that confronted her, McClintock left Brooklyn for her new home in Ithaca, New York as a student at Cornell University for the fall semester in 1918.

Cornell was then really two schools. It was comprised of a private liberal arts college and a state-funded agriculture ("ag") school. Students at either school could take any course the other school offered. Even though she loved the science and math that she had taken in high school, McClintock had no

Barbara McClintock's graduation photo from 1923, the year she completed her bachelor's degree in plant breeding and botany. She was doing graduate work before her official graduation, and an invitation from a professor who saw potential in her would decide the course of her life.

apparent inclination towards agriculture. Almost all of the students in the ag school meant to become farmers. It is probable that she began her studies in agriculture simply because it allowed her to take other courses, like meteorology and music, that interested her more. Gradually, however, she would become involved in one branch of science that deeply interested agriculturalists. Unknowingly she became immersed in the most exciting and up-to-the-minute field of her time, the infant field of genetics. New and exciting discoveries in genetics and heredity were being published almost daily.

The area of study now referred to as genetics had begun about 50 years earlier with the work of an obscure Czechoslovakian priest, Gregor Mendel.

GREGOR MENDEL

Gregor Johann Mendel was born on July 22, 1822 in Hyncice, Moravia—in what is now the Czech Republic. He was the son of a poor farmer and attended the local schools. He was ordained a priest on August 6, 1847 and thought that he was destined to become a teacher in the Augustinian Order of Monks. But he failed his first examination to receive credentials to become a teacher. The Augustinian Order sent Father Mendel to Vienna for two years, where he attended classes in the natural sciences and mathematics in order to prepare for his second try at the state examination. It was in Vienna that he learned the skills he would later need to conduct the experiments that would make him famous. He never passed the teacher's exam—he was so afraid of failing again that he became ill on the day of the exam, and the Augustinians gave up on Father Mendel's ever becoming a teacher.

Left to himself much of the time, Mendel began his scientific experiments after his return from Vienna. His research involved careful planning, necessitated the use of thousands of experimental plants, and, by his own account, extended over eight years. Prior to Mendel, heredity was regarded as a "blending" process and the traits of the offspring as essentially a "dilution" of those of the parents. Mendel's experiments showed that this was not so.

He took two sets of pea plants—one set that had been bred for many generations to produce only peas with smooth skins and another set that had been bred for many generations to produce only peas with wrinkled skins—and he mated them. If inherited characteristics did actually blend, he knew, the offspring would have peas with a little wrinkling.

But the first generation of offspring of that original cross all had smooth skins. What had happened to the inherited characteristic of wrinkled skin? When Mendel crossbred members of that first generation, he found that some of the second generation of peas *were* wrinkled. In fact, almost exactly 25% of that generation of peas had wrinkled skins. The mechanism that created those wrinkled skins clearly had not been lost or destroyed in that first generation. Mendel reasoned that somehow the characteristic for smooth skins (designated S) *dominated* the characteristic for wrinkled skins (designated s), which he called a *recessive* characteristic.

The almost exact percentage of 25% wrinkled peas exploded with brilliant clarity in Mendel's mind. Mendel saw that every reproductive cell, or gamete, of each parent pea must contain *two* specifiers of a given characteristic, such as skin type—for peas, smooth (S) or wrinkled (s). One kind of specifier, or gene, must be dominant and one recessive, so one characteristic always will be more likely to be expressed in the offspring than the other. (Eye color in humans is an example: brown is dominant and blue recessive, so a child who has genes for both usually will have brown eyes.) The next-generation pea would again contain two copies of a characteristic—one donated by the father and one by the mother. A cross between a purebred smooth pea (S, the dominant characteristic) and a wrinkled pea (s, the recessive characteristic) would always result in a smooth-skinned pea (S), for in peas the gene for smooth skin is dominant and the gene for wrinkled skin recessive; but each of them would retain one copy of the recessive characteristic(s).

When first-generation peas are crossed, both the sperm cells and the egg cells can contain either characteristic, smooth (S) or wrinkled (s). When the next generation is created, three of every four will contain at least one copy of the dominant gene (S) and will produce smooth-skinned peas. One of the four will contain two copies of the recessive gene for wrinkled skin (s) and produce peas with a wrinkled skin, since there will be no copy of the dominant (S) gene within the organism. Only in the absence of the dominant gene will the recessive gene express itself.

It was a brilliant insight, which must have come entirely to Mendel's mind in a single vision, in what is called an intuition or intuitive leap. Throughout the writings of great minds who have made quantum leaps in man's progress in understanding nature, people like Einstein and Galileo, the writers say that their greatest ideas and theories came to them in such intuitive leaps.

Mendel presented his work in a series of two lectures before the Society for the Study of the Natural Sciences in 1865. He published the work as "Versuche über Pflanzen-Hybriden" ("Experiments in Plant Hybridization") in the society's *Proceedings* in 1866. The Society sent 133 copies to libraries and other scientific societies around the world. Mendel paid for another 40 copies, which he sent to friends. His work was largely ignored. Mendel died in Brünn on January 6, 1884. Just before his death he commented, "My scientific labors have brought me a great deal of satisfaction, and I am convinced that before long the entire world will praise the result of these labors." His foresight proved as true as his scientific vision—today he is regarded as one of the great biologists of the 19th century. In the spring of 1900, three botanists, Hugo de Vries of Holland, Karl Correns of Germany, and Erich von Tschermak of Austria, reported independent verifications of Mendel's work which amounted to a rediscovery of his work. They all, knowing nothing of Mendel's work, came to the same results through their own independent experiments.

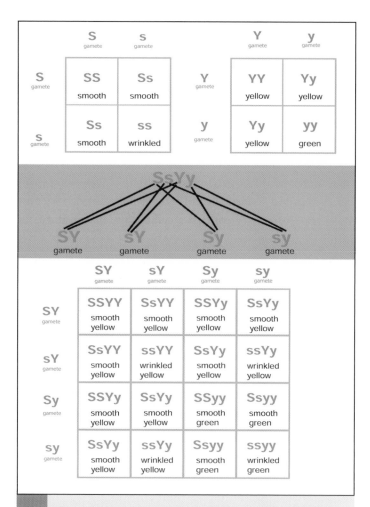

Punnett squares (named for the British geneticist Reginald Punnett) show the chances that certain genetic combinations will result from the crossing (mating) of certain others. The above examples illustrate Gregor Mendel's early work with peas: the genes for smooth and yellow skin are dominant, and the genes for wrinkled and green skin are recessive. The 2-by-2 squares above are called *monohybrid crosses*, as they involve only one trait; the 4-by-4 square below illustrates a *dihybrid cross*, demonstrating the probability that given combinations of skin type and skin color will be produced in the offspring of the parent plants.

Mendel never attempted to find the units inside the pea that were responsible for inherited characteristics, but that became the immediate goal of many who followed quickly behind him.

CHROMOSOMES

Within a few years, scientists had settled on the chromosomes inside the cell nucleus as the site for the inherited characteristics that are passed from parents to children. Chromosomes had been discovered accidentally in 1847, when the German scientist Wilhelm Hofmeister colored a cell with a chemical dye in order to make the tiny, almost transparent parts inside a cell visible under a microscope. Hofmeister discovered that when cells were in the process of dividing, tiny rod-like structures appeared in pairs. These apparently were divided among the new "daughter" cells during reproduction. These "chromosomes" appeared in every plant and animal that was studied. The word *chromosome* was proposed by Heinreich Waldeyer in 1888; it combined the Greek words for "colored" (*chromo*) and "body" (*somos*). Because of Mendel's work, these chromosomes were suspected of being the location for the inherited characteristics. In 1891, Hermann Henking demonstrated that indeed reproduction within the cells began with the conjugation of chromosomes, two by two.

Johannes Rückert suggested in 1892 that in sexual reproduction one chromosome in a pair came from each parent, and that they could exchange material and thus create chromosomes with parental characters in new combinations. Over the next few years, Thomas H. Montgomery and Walter Stanborough Sutton confirmed Rückert's theory. Sutton studied the eleven pairs of chromosomes in grasshopper cells and concluded that while all of the pairs differed in size, the members of a pair were the same size. He studied the chromosomes during reproduction and concluded, "[T]he association of paternal and maternal chromosomes in pairs and then subsequent separation during the reducing division [later known as *meiosis*] may indicate the physical basis of the Mendelian law of heredity." He published this in 1903, when he was 23 years old. All of this work laid the foundation for the next major leap in the study of heredity— the work of T.H. Morgan.

T.H. MORGAN

Thomas Hunt Morgan was born on September 25, 1866 in Lexington, Kentucky. He earned a bachelor's degree at the University of Kentucky in 1886. As a postgraduate student he studied morphology with W.K. Brooks and physiology with H. Newell Martin. In 1891, Morgan became an associate

AGRICULTURAL SCHOOLS IN THE U.S.

When Barbara McClintock enrolled at Cornell, American colleges had already been offering classes in agriculture for almost one hundred years, since 1825. In 1855, Michigan founded the first agricultural college—the first institution that existed solely to educate future farmers. Under a new federal law, many more states opened what are sometimes called "ag" schools. In the 1870s, these schools started to do experimental work, such as testing various planting techniques. Over time, they would be credited with learning a great deal that directly helped farmers. In 1920, there were 31,000 students enrolled in the nation's agricultural colleges. The government was so pleased with the work these universities did that it gave them money to expand their programs. By 1940, there were 584,000 students enrolled in the nation's agricultural colleges. Today many of these schools offer a broad range of courses. They continue to be important places for research, including a great deal of key scientific work in botany and zoology.

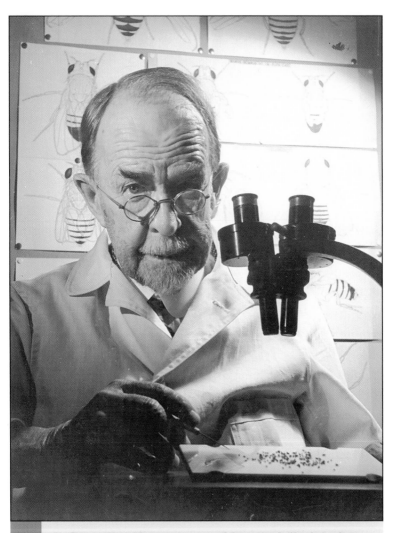

Dr. Thomas Hunt Morgan was a geneticist at the California Institute of Technology, where he discovered that some genetic traits were linked to gender. He was also the first to show that physical objects on chromosomes—later called *genes*—were real. It was because of Morgan's compelling recommendation to the Rockefeller Foundation that McClintock received a grant to continue her research after she left Cornell; it was also Morgan who urged McClintock and Creighton to publish their important early paper on genetic crossover in 1931.

professor of biology at Bryn Mawr College for Women, where he stayed until 1904, when he became a professor of Experimental Zoology at Columbia University in New York. He was a passionate research scientist, and he passed his enthusiasm on to his students. He worked with fruit flies in his heredity experiments; his 16' x 23' laboratory, known as the "fly room," was filled with active student workers, milk bottles containing buzzing fruit flies (millions of them), and the smell of the bananas that he fed to the flies.

Morgan found that some inherited characteristics in his fruit flies appeared together more frequently than could be predicted from strict Mendelian rules. That is, some characteristics seemed to be linked, as though they were somehow tied together. In May of 1910, Morgan discovered that one of the eyes of a male fruit fly was white instead of red as in all the other fruit flies he had. Morgan crossed the white-eyed male with a red-eyed female and got white-eyed males and red-eyed females. (The characteristic of red eyes is dominant over that of white eyes.) Because the white-eyed trait appeared only in males, he referred to it as a "sex-limited" characteristic. He assumed that the characteristic was contained within the sex characteristic.

However, he then mated the original male with some of the red-eyed daughters, and he eventually obtained white-eyed daughters, each of which had two copies of the recessive white-eyed gene. Thus, the characteristic of white eyes was not carried in the trait that determines sex of the offspring, but *was* carried on the same chromosome as the one for gender. It was a "sex-linked" characteristic. Each chromosome was responsible for more than one character trait, so each character trait must be a unit smaller than the chromosome; in other words, each chromosome must comprise several of these character units. Morgan later found another abnormal characteristic—yellow-bodied fruit flies. The characteristic for yellow bodies behaved exactly as did the one for white eyes. Both were tied to the chromosome that determines gender; both were sex-linked characteristics. In this work, Morgan provided the first evidence that the units of inherited characteristics are real, physical objects, located on chromosomes, with properties that can be manipulated and studied experimentally. It was found that genes in chromosomes normally occupy the same fixed positions on the chromosomes relative to each other, but this was later found not always to be true.

Wilhelm Johannsen proposed the term *gene* for this character-bearing unit within the chromosome in 1911. He explained that *gene* was "nothing but a very applicable little word, easily combined with others, and hence it may be useful as the expression for the 'unit-factors' demonstrated by modern Mendelian researches." (Johannsen, 1911) They were originally imagined to be the ultimate causes of inherited characteristics or the mutations (spontaneous changes) of them. The white-eyed fly provided the foundation upon which Morgan and his students would establish the modern theory of the gene.

"Crossing over" (see page 54) was another famous discovery made in the "fly room" of T.H. Morgan and his graduate students. Morgan discovered that some of his crosses between fruit flies simply did not follow the predicted percentages of characteristics that would come from Mendel's laws. Sometimes, flies were produced that had only one or a few of the expected sex-linked characteristics. They would occasionally produce yellow-bodied flies with red eyes, for example, when usually yellow bodies and white eyes occurred together (a result of being on the same chromosome). In these aberrant flies, a part of the sex-linked chromosome must have "crossed over" to another chromosome or become lost or destroyed. Morgan hypothesized that the less frequently such a crossover occurred, the closer the genes were on the chromosome. By this theory, it would be possible to construct a map of where on the chromosome each of the genes was located. The more frequently two given traits were expressed together, the closer together their genes must be.

In 1915, Morgan and his students published all their findings in a book entitled *The Mechanism of Mendelian Heredity.* This book provided the foundation for modern genetic theory. McClintock probably either found it in the agricultural library at Cornell or read it as an assignment in class. For his discoveries concerning the role played by the chromosome in heredity, Morgan won the Nobel Prize in 1934.

MCCLINTOCK AT CORNELL

During her first few semesters at Cornell, McClintock remained quite unconcerned with the advances being made in genetics, because she was more interested in the freedom that college life offered. Her lively personality blossomed in the college atmosphere. She was popular, and she was respected enough to be elected president of her freshman class. She was approached to join a sorority, too. At first, she was delighted by the invitation, but when she learned that sororities were very exclusive and that she was the only girl in her rooming house to be chosen, she declined to join:

> Many of these girls [sorority members] were very nice girls, but I was immediately aware that there were those who made it and those who didn't. Here was a dividing line that put you in one category or the other. And I couldn't take it. So I thought about it for a while, and broke my pledge, remaining independent the rest of the time. I just couldn't stand that kind of discrimination. (Keller, 33)

When the soldiers returned from Europe after the end of World War I, they brought with them a new attitude toward life. They rejected the social conventions of their parents. The especially severe horrors of trench warfare and the general loss of faith in progress made these young men desperate

Cornell University in the Roaring Twenties, when McClintock was a student there. The 1920s was a period of dramatic change in society, as the soldiers who had returned from World War I were less inclined to follow the strict morality of their parents. McClintock caught this spirit as well, taking any class that pleased her and not caring much about her grade point average; she was also among the first at Cornell to "bob" her hair.

to enjoy the pleasures of life without the moral constraints that inhibited their parents. A new age dawned, of wild music, jazz, loose sexual mores, and drinking and dancing in speakeasies. Today this decade is remembered as "the Roaring Twenties." Things were changing especially quickly for women, too. When McClintock was born, American women did not have the right to vote. That right was finally granted—thanks mainly to women like Lucy Burns, one of McClintock's teachers at Erasmus—in 1920, with the ratification of the 19th Amendment to the United States Constitution.

The unconventional, independent, self-reliant McClintock took to the new times like a fish in water. Before any other girl on campus, she had her long hair cut into a layered pageboy, "bobbed," style. She smoked cigarettes in public. She not only listened to jazz but played it. Although she loved music, she would never develop a great musical talent; but this did not stop her from playing the tenor banjo in a local jazz band in the bars and restaurants in downtown Ithaca.

She enjoyed school and expressed at least an initial interest in almost every course offered. She often began a course and quickly dropped it when it proved dull. Each time she did this, she received a "Z" on her record, which counted against her grade point average. But a high grade point average was not something McClintock was much interested in:

> At no time had I ever felt that I was required to continue something, or that I was dedicated to some particular endeavor, I remember I was doing what I wanted to do, and there was absolutely no thought of a career. I was just having a marvelous time. (Keller, 34)

By the beginning of her junior year, McClintock found herself well on the way to a degree in cytology, the study of the mechanisms of cells. In the fall of 1921, she attended the only genetics course open to undergraduates at Cornell University, taught by C.B. Hutchison, a professor in the Department of

Plant Breeding. It was a small class with only a few students, most of whom wanted to go into agriculture as a profession. That year Hutchinson published an article entitled "The Relative Frequency of Crossing Over in Microspore and in Megaspore Development in Maize" in the influential journal *Genetics*. He expressed a particular interest in McClintock.

This gracious interest by a distinguished professor impressed McClintock deeply, as she recalled in her Nobel autobiography in 1983:

> When the undergraduate genetics course was completed in January 1922, I received a telephone call from Dr. Hutchison. He must have sensed my intense interest in the content of his course because the purpose of his call was to invite me to participate in the only other genetics course given at Cornell. It was scheduled for graduate students. His invitation was accepted with pleasure and great anticipations. Obviously, this telephone call cast the die for my future. I remained with genetics thereafter.
>
> At the time I was taking the undergraduate genetics course, I was enrolled in a cytology course given by Lester W. Sharp of the Department of Botany. His interests focused on the structure of chromosomes and their behaviors at mitosis and meiosis. Chromosomes then became a source of fascination as they were known to be the bearers of "heritable factor." By the time of graduation, I had no doubts about the direction I wished to follow for an advanced degree. It would involve chromosomes and their genetic content and expressions, in short, cytogenetics. This field had just begun to reveal its potentials. I have pursued it ever since and with as much pleasure over the years as I had experienced in my undergraduate days.

McClintock was skilled at using the microscope, and she often could see structures and changes within the cell that others could not see. This skill gave her an edge in her future research and made her successful where others had failed.

In June of 1923, McClintock received her undergraduate degree. At that time, approximately 25% of the graduates from the College of Agriculture were women. McClintock's undergraduate majors were Plant Breeding and Botany, in which over the years she had earned a grade point average just under a B. There was not yet a degree offered in genetics, because the field was still too new. On invitation from Dr. Hutchinson, her undergraduate genetics professor, she opted to stay at Cornell for graduate study. McClintock enrolled as a equivalent to doctoral student in the Department of Botany. She majored in cytology with minors in genetics and zoology.

PASSION FOR THE STUDY OF MAIZE

McClintock decided to focus her graduate research on the cytology and the genetics of maize—how its cells are formed and structured, how they function, and how they pass on their genes. There was a professor there, Rollins Emerson, who was then a leader in the field, which meant she had a nourishing environment in which to study. Maize is a common form of corn, often called "Indian corn"—though in fact the two words, one from Europe and one from the Taino people of San Salvador, both mean "source of life" and really refer to the same plant. Its scientific name is *Zea mays*, *zea* being a Greek translation of the same "source of life." Scientists chose maize for genetic studies because its multicolored kernels made it easy to keep track of which dominant and recessive traits were being passed from one generation to the next. While the genetics departments of other universities were studying *Drosophila melanogaster* (the fruit fly), the Cornell group concentrated on maize.

Rollins Emerson taught McClintock and all the other budding geneticists at Cornell how to grow corn so carefully that they could be sure of the heritage of each plant. Corn is self-pollinating—that is, both the male and female reproductive

Dr. Rollins Emerson was McClintock's mentor during her graduate study at Cornell. Emerson was a leading expert in plant cytology and the study of maize. It was Emerson who showed McClintock how to raise maize plants in such a way that they would not cross-pollinate with other plants, thus making it possible to study the effects of certain genes on specific plants.

organs are on the same plant. The male reproductive organ of the corn plant is the anthers that comprise the tassel at the top of the plant. Tiny silk threads emerge from the nascent ear along the stem of the plant when the female parts are ready to be pollinated. These threads capture the male pollen and direct it downward to be fertilized. Before this occurs, when the stalks are still very small and hard, the researchers must cover each of the shoots with a transparent bag and tie it securely. The stalk then grows inside the bag, which allows its receive sunlight and be observed but at the same time prevents it from being pollinated by nearby plants (by protecting it from any pollen that may be in the air). Pollination by the wrong plant would ruin the experiment. This process is called "shoot-bagging."

Each day McClintock would tend her plants, weeding, watering, and just watching. Early on the morning when McClintock had decided to make the fertilization, she would go to the field and strip the anthers from the tassel of the plant and then put a brown paper bag over the top of the tassel, again fastening it securely. The male pollen is viable for only a few hours—the first anthers she collected were at least one night old and not positively viable. The tassels produced pollen all the time during the morning, so by bagging the male part, McClintock could be assured of collecting more pollen later in the day. After breakfast, McClintock would return to her plants, strip the anthers while they were still inside the brown bag, and carefully pour the saffron-colored pollen onto the silk threads that she had just exposed to the air. The silk threads of the female are sticky, so the pollen easily covered and remained on the threads, giving them a fuzzy yellow appearance. McClintock would then replace the transparent bag over the now impregnated threads and let nature take its course.

McClintock would note the fertilization on a stick that she would place in the ground next to the plant and then make a much more detailed index card, which she would store in her records in the laboratory.

Only one maize crop a year could be grown in the climate of New York, and that crop was subject to all the chance happenings of nature. Crows often ate some of the crop. Drought and floods came and destroyed McClintock's tiny field just as they did the thousand-acre farm nearby. As the crop matured, the mutations and chromosomes each plant carried would become obvious in differently colored leaves, stunted growth, or any of dozens of other characteristics. After the growing season, McClintock would carefully select the kernels for the next year's planting. She spent the winter months in the laboratory, at her microscope, peering deeply into the wonderful mysteries of tiny rod-like stains on the carefully prepared slides below. She recorded the information she collected and used her data to formulate theories as to why what she saw occurred. Thus in studying genetics she mastered three separate sets of skills: she became a savvy farmer, an expert at conducting experiments, and a talented theoretician. Other scientists who used her studies would comment on how rigorous her work was and praise her painstaking research.

McClintock worked on her graduate research with her characteristic independence and passionate drive. In her first year as a graduate student, she discovered a way to identify maize chromosomes. McClintock's skill with the microscope allowed her to distinguish the individual members of the set of chromosomes within each cell—a startling and incredible achievement for someone so early in her career. She found that maize has ten chromosomes, in five pairs, in each cell. She was able to give each maize chromosome a label and an identity so that she could follow it through its life cycle. Each chromosome has its own length, shape, and structure, which is known as the chromosome's *morphology*. She came to know the look of each of those chromosomes—the arrangement of the "knobs" on its surface. Once McClintock had determined the correct number of chromosomes in maize, she turned her attention to determing which genes were contained in each of those chromosomes. This was a far more daunting task.

Early in her research, she developed "a feeling for the organism" that allowed her to see the smallest microscopic changes, unseen by others before her. (Keller, xiv) These microscopic differences in the cell revealed themselves as differences in the adult ear of corn. Differences such as the colors of the kernels could be linked to differences in the individual chromosomes. This may seem simple to scientists now, but in the mid-1920s McClintock's discovery was groundbreaking. She published papers based on her findings and completed her graduate thesis. She had received a master's degree in 1925. Two years later, in 1927, approaching her 25th birthday, McClintock received her Ph.D. in botany from Cornell.

4

Choosing a Career: 1927–1941

It might seem unfair to reward a person for having so much pleasure over the years, asking the maize plant to solve specific problems and then watching its responses.
—Barbara McClintock, 1983

McClintock's qualifications were outstanding, and her potential was obvious to those who knew her at Cornell. She was offered the position of Instructor in Botany at Cornell, which she accepted. This position would allow her to continue with her maize research, which was her main concern. She was soon surrounded by other brilliant scientists, each of whom fed on the enthusiasm of the others and all of whom shared her interests. They included George W. Beadle, who had come to start work on a Ph.D. in plant breeding. In the fall of 1928, Marcus M. Rhoades came on board. He already had a master's degree from the prestigious California Institute of Technology (Cal Tech), but,

McClintock at the Marine Biological Laboratory (MBL) at Woods Hole, Massachusetts, in 1927. She studied botany at the laboratory there for a short time while working as an Instructor in Botany at Cornell. Her self-reliance seems evident, even this early in her career.

like Beadle, he had come to do his doctoral work under Emerson. He knew all about the work being done at Cal Tech with *Drosophila*. This group greatly enjoyed discussing how chromosomes and genes worked and welcomed input from any new graduate student.

One of McClintock's first research collaborators was

another woman, Harriet Creighton, who arrived at Cornell in the summer of 1929. Creighton had just graduated from Wellesley College, a women's college near Boston, Massachusetts. Many women earned undergraduate degrees in botany from Wellesley, and many went on to graduate studies at Cornell and the University of Wisconsin. Creighton and McClintock met on Creighton's first day at Cornell. McClintock took charge of Creighton and introduced her to her former advisor, Lester Sharp. Creighton decided to follow McClintock's suggestion and major in cytology and genetics.

Working with the maize was a demanding job. Before she was able to use her microscope in the laboratory, McClintock had to plant, grow, observe, and harvest the maize. She had to extend the growing season of the corn for as long as possible. This meant choosing the warmest place in the field to plant the seeds and then making sure the plants got the right amount of water and didn't dry out.

The days were long and physically demanding, but McClintock enjoyed them. She even had the energy to unwind by playing tennis with Creighton at the end of every day. McClintock was a determined, gutsy player. She went after every ball—she put maximal effort into every play.

Harriet Creighton later told a story about McClintock at this time. One June, they had a project going together. The corn plants they were growing stood only about a foot high when a "once-in-a-century" torrential rain struck Ithaca. It lasted for hours, all night long. Creighton had to get up at 3:00 in the morning to help her mother evacuate her house because a nearby creek had risen above its banks. After all her mother was safe, Creighton drove her car to the fields where the geneticists grew their corn. There she found McClintock working in her corn plot, building up the soil around the plants that remained in the ground. Some of the plants had been washed away; in some cases she could tell where they had come from and replant them in their assigned places. All of the plants had

THE MARINE BIOLOGICAL
LABORATORY AT WOODS HOLE

In 1927, while she was an instructor at Cornell, McClintock studied botany at the Marine Biological Laboratory (MBL) in Woods Hole, Massachusetts. The MBL, founded in 1888, had a long history of supporting the work of women. At a time when science was dominated by men and women had to struggle even to be taken seriously—McClintock certainly faced this problem in her early days at Cornell—the MBL took the courageous stand of opening its enrollment to all qualified applicants of *both* sexes. From the institution's founding until 1910, women accounted for approximately one third of the total enrollment. The number of female applicants decreased after that, and it remained low until the 1970s. Woods Hole has never granted degrees to its students, but many of the women who have studied there have gone on to earn doctoral degrees at other institutions of higher learning. The equal footing of the sexes at Woods Hole, as well as its enormous library, attracted some of the most illustrious biologists in the world.

McClintock is only one of the many students from the MBL who went on to achieve great things. The author and ecologist Rachel Carson, also profiled in this series, explored her love of biology there before she became a graduate student. Gertrude Stein, one of the great innovators of 20th-century literature, studied embryology at Woods Hole just after her graduation from Radcliffe College, when she was considering a career in medicine. Woods Hole boasts 37 affiliated winners of the Nobel Prize, among them McClintock's colleagues George Beadle and Thomas Hunt Morgan, and the MBL continues to attract talented scientists from all over the world.

been numbered, and a small sample of the root tips had been taken in order to determine the number of chromosomes in each plant. "If she had lost them, she would have blamed only herself, not nature or fate, for not having made the maximum effort to save her research." (Fedoroff and Botstein, 14)

CROSSING OVER

What McClintock and Creighton wanted was to show that "crossing over" happened in the chromosomes of maize. Crossing over is a physical exchange of segments of a chromosome that occurs after it has replicated itself during the first phase of meiosis (one of the ways in which a cell replicates itself). For some reason two chromosomes lying close to each other can exchange pieces of chromosomal material. Each breaks along its length and joins with the other part, and usually the break happens at different points on the two chromosomes.

Having found evidence of crossing over, the women wrote up their results and published them in a paper entitled "A Correlation of Cytological and Genetical Crossing-Over in *Zea Mays*" in the prestigious journal *Proceedings of the National Academy of Sciences* in August of 1931. Creighton and McClintock had demonstrated that genetic crossing over was accompanied by physical crossing over of the chromosomes—a milestone in the study of genetics.

A PRODUCTIVE TIME

McClintock was very productive in the early years of her career. She published nine papers between 1929 and 1931, and each of these made a major contribution to the understanding of the structure and genetic markers of maize. She also continued to spend time with her colleagues Marcus Rhoades and George Beadle. The three researchers were young and confident pioneers in a new and exciting field, and their work was already receiving recognition in the scientific community. For Rhoades and Beadle, the path was clear: young men of their brilliance and

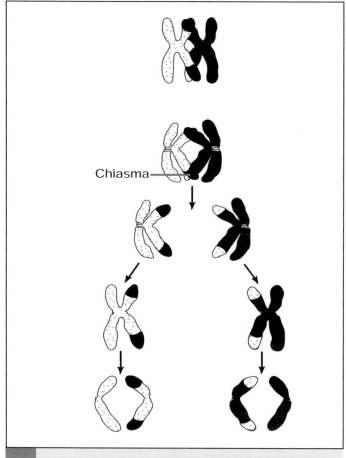

Chiasma

Genetic "crossing over" occurs during the division of reproductive cells (meiosis). The legs of chromosome pairs touch at the *chiasmata* (sing. *chiasma*), and then the chromosomes exchange segments and move apart. It is through this crossing over, which creates single chromosomes with new combinations of genes, that a human child inherits some of its mother's traits and some of its father's.

ambition were expected to become professors within the American university system. But for McClintock the path was neither so straight nor so easy: she remained as an instructor at Cornell, but by 1931 she believed the time had come for her to leave.

The stock market crash of 1929 had started the period of the Great Depression in the United States, and jobs were not easy to find. But McClintock managed to garner a fellowship from the National Research Council, which would support her financially for two more years; this grant enabled her to travel among the University of Missouri, the Cal Tech, and Cornell. In each place, she continued her research on maize.

CHROMOSOMAL RINGS

X-rays were discovered in 1895. They were almost immediately used to treat various human illnesses, such as tuberculosis to tonsillitis. By 1902, the first cases of cancer that were associated with the use of X-rays were reported in patients in Germany and the United States. Tumors were often found on laboratory workers who used X-rays, too. In 1908, a scientist in Paris exposed four white mice to large amounts of radiation; two died almost immediately, and one developed a large cancerous growth at the site of the irradiation. In 1927, Herman Muller, working at Columbia University with T.H. Morgan, found that he could induce mutations, permanent changes in the genetic material, in fruit flies. Muller's mutant fruit flies came in all shapes and sizes—big eyes, no eyes, hairy bodies, bald bodies, and short- and long-lived flies. From a theoretical standpoint, Muller's work demonstrated that physical agents could alter genes—which implied that genes had a definite normal structure that could be changed. (Muller was awarded the Nobel Prize for this work in 1946.)

Lewis Stadler, one of the premier maize geneticists in the country, also became interested in the effects that X-rays had on genes at about the same time that Muller was doing his Nobel Prize–winning experiments. Stadler and McClintock had become acquainted in 1926 when he'd worked at Cornell on a research fellowship. Stadler began irradiating corn with X-rays in order to produce mutations, much as Herman Muller had done. The corn that was the result of these X-ray experiments were displaying strange characteristics, most noticeably in the color and texture of the kernels. Ears of corn were being produced that had ten or twelve different-colored kernels. Stadler began sending samples of these kernels to McClintock at Cornell for her to grow and examine. Stadler wanted McClintock, with her keen eyesight, to identify the kinds of mutation he was getting in the X-rayed corn. McClintock quickly realized that irradiating corn chromosomes caused them to break in unexpected places. She saw all kinds of damaged chromosomes. McClintock later remembered, "That was a profitable summer for me! I was very excited about what I was seeing, because many of these were new things. It was also helping to place different genes on different chromosomes—it was a very fast way to do it." (Keller, 65)

That fall, McClintock received a reprint of an article from Cal Tech in which some of these chromosomal abnormalities were explained by the hypothesis that a part of the chromosome had broken off completely from the parent chromosome and, because its ends had fused together to form a stable ring, had become incapable of changing further in the process of cell division. When McClintock read this, she realized immediately that some of the changes in Stadler's irradiated Missouri corn were the result of these "ring chromosomes."

McClintock continued her investigation into the nature of ring chromosomes. It quickly became obvious to her that the broken ends of chromosomes were incredibly reactive—that they seemed desperate to link to other pieces. Once they were broken, the chromosomal fragments quickly joined any broken chromosomal end nearby, even the other end of their own broken chromosome, in which case they would form a ring. The chromosome from which the ring was formed then quickly "healed" itself, closing off all possibility of further interaction with the ring, even if the ring were capable of it. The original chromosome, called a *deletion* because it had lost some of its genetic material permanently, would then continue the process of reproduction, and the material that had been lost would be lost to any offspring. Often, this didn't mean too much, as the missing genes would be expressed by the chromosome obtained from the other parent; but it did mean that the offspring had only one chance, not two, to inherit any normal characteristics controlled by those missing genes.

It seemed to McClintock that this had to be a natural function of the wonders of reproduction. Normal sexual reproduction was designed to allow the offspring the best chance of survival regardless of the behavior of the chromosomes during reproduction. If one chromosome were damaged, the offspring had another chance given to it by the other parent. Thus too, although X-rays created strange and damaged chromosomes in great multiplications from the norm, it must be a normal, and for some reason desirable, function of chromosomes to break, reform, and reproduce in unexpected

MUTATIONS

Mutations are a form of adaptation to the Earth's constantly changing environment. Some of the beneficial mutations that occur in DNA allow organisms to adapt to this changing environment, and without this ability to change species would become extinct. However, most mutations are harmful and cause many of the genetic diseases that are discovered by researchers today. Human fetuses are very susceptible to mutations while they are in the womb. The exposure of the mother to X-rays, tobacco smoke, drugs, alcohol, and other chemicals can cause mutations to the genes that damage the fetus and result in deformed babies or stillbirths. These toxic elements are called *teratogens*. A miscarriage during pregnancy may be nature's way of ensuring that a baby whose mutations are too extensive for its survival is not born.

ways. Life would try to express its variations through these chance changes in chromosomes. Because chromosomes were expressed in pairs, many of these abnormal chromosomes would never be expressed; but every once in a while, the viability of an offspring would depend on the new chromosomes. It would be a form of evolution.

McClintock wrote to Stadler of her findings. She told him that she was very enthusiastic about his X-ray techniques and asked him to grow more specimens the next year for her to work on. He was happy to do this and invited her to Columbia to examine them herself; she accepted, and when she went, in the following summer, she would be surprised—and a little worried—to see that everyone there, while kidding her about her obsession with ring chromosomes, had already labeled the plants as a ring crop.

Before she went to Missouri in the following summer, though, McClintock used the money that had been awarded to her with a 1933 Guggenheim Research Fellowship to visit her friends at Cal Tech and make a trip to Germany. Like most of her peers, McClintock did use the Guggenheim to continue her studies—but she made sure to choose faraway and interesting places in which to conduct those studies. Germany had been a hub of scientific activity since the early part of the 19th century. Because most scientific journals were written in German, every undergraduate science student took German classes in college. It was not important to be able to speak German, but it was essential to be able to read it well. McClintock looked forward to being a tourist in Germany, where she would meet and work with some of the most famous scientists in the world.

DISAPPOINTMENT IN GERMANY

In 1933, McClintock was awarded the Guggenheim fellowship that took her to Germany. Morgan, Emerson, and Stadler recommended McClintock for this prestigious award, and she greatly appreciated the opportunity, but her time in Germany

Curt Stern was a prominent geneticist in Germany; he studied *Drosophila*, or fruit flies, instead of maize. McClintock had traveled to Germany in 1933 to meet Stern, but by then he had already fled the country because he was Jewish and subject to Nazi persecution.

proved traumatic for her. Germany was already under the influence of Adolf Hitler's Nazi regime. McClintock planned to study with *Drosophila* geneticist Curt Stern— whose research McClintock and Creighton had "one-upped" two years earlier—but Stern had already fled Germany because he was Jewish and therefore in danger from the rising regime. On March 23, 1933, the German imperial parliament, the Reichstag, had passed the Enabling Act, giving Hitler dictatorial power over the nation. The climate was one of repression and persecution. The press came under Nazi control, and the books of "undesirable" authors were burned. Educational institutions and the young people attending them were also under strict supervision and control to prohibit the fostering and expression of dissident views. Nazi ideology became the basis of national law, and enforced Nazi rules replaced former legal procedure. The strict Nuremberg Laws forbade intermarriage with Jews, deprived Jews of civil rights, and barred them from certain professions. Similar laws were enacted prohibiting Communists from living freely. Hitler's special police forces, the S.S. and the Gestapo, were beginning their reign of terror. Even the wealthy and well-connected were not safe if their beliefs did not conform. People were taken from their homes by the Gestapo in the middle of the night and never heard from again.

The combined persecution of the Jews and Communists and those involved in education quickly took its toll on Germany's scientific community. When McClintock arrived in Berlin, she found that few people were left at the Kaiser Wilhelm Institute and practically no foreigners like her. During the two months she stayed in Berlin, she felt both physically ill and mentally discouraged. In addition, she had difficulty with the language, which must have added to her sense of isolation and despair.

But, fortunately, she met Richard B. Goldschmidt, the head of the Institute and an important geneticist in his own right. He suggested she leave the Institute to do research at the Botanical Institute in Freiburg. He himself was heading there; he had a son and daughter to worry about, and he hoped the move would ensure their safety—as well as his own. So far, Goldschmidt's prominence had protected him, but everyone was beginning to understand that there were no guarantees for anyone under the Nazi regime.

Freiburg was a small university town, which McClintock found beautiful and easy to live in. In a letter to Curt Stern and his wife, she described the Botanical Institute as "well equipped although there are relatively few people here now." There were a number of Americans still in Freiburg with whom McClintock was able to exchange ideas and questions. But even though Freiburg was more hospitable than Berlin and Goldschmidt continued to support her in her studies, McClintock remained miserable in Germany. Goldschmidt himself was not going to remain in Germany for long: in 1936, after the passage of even more racist legislation, he left for the United States, where he joined the Department of Zoology at the University of California at Berkeley. (He later said that moving to the United States, where he became an American citizen, in 1942, was one of the happiest things ever to occur for him.) He continued his research and taught both genetics and cytology for more than a decade.

McClintock returned to Cornell in April of 1934, dispirited from her experience in Germany. Nonetheless, she readily admitted to having learned a great deal while traveling abroad. She'd seen firsthand the destruction caused by Hitler's anti-Semitism and feared for her Jewish friends and colleagues. Later that year she wrote to Stern about her trip: "I couldn't have picked a worse time. The general morale of the scientific worker was anything but encouraging. There were almost no students from other countries. The political situation and its devastating results were too prominent." McClintock returned home to Cornell without a plan for her future.

Around this time, McClintock came to understand herself as a "career woman." She later recalled, "[In] the mid-thirties a career for women did not receive very much approbation. You were stigmatizing yourself by being a spinster and a career

When McClintock was there, Cornell was divided into two schools: a liberal arts college and an agricultural school. This is the Plant Science Building, built in 1931, where the Department of Botany has been based for decades. The Department was based at first on the third floor of Stone Hall, just to the west of this building, but it took up residence on the first floor of Plant Science not long after the building's construction. McClintock's office in Plant Science was on the floor above.

woman, especially in science. And I suddenly realized that I had gotten myself into this position without recognizing that that was where I was going." (Keller, 72) But she had few if any regrets about the unwitting choice she had made. She would always find her work absorbing and fulfilling.

THE NEED FOR FUNDING

McClintock confronted the fact that she needed a stipend to support her; otherwise, she would be unable to continue her research. In the spring of 1934, the country still reeled from the collapse of the stock market five years earlier. The era known as the Great Depression had not yet come to an end. All over the country, people were out of work. Opportunities were scarce, even for someone with McClintock's excellent qualifications. But her friends in the scientific community came to her aid: When Rollins Emerson learned she did not want to come back to Cornell, he contacted T.H. Morgan at Cal Tech. Morgan in turn contacted the Rockefeller Foundation and asked that it award to McClintock a grant to support her research in maize genetics. When contacted, Emerson convinced the reviewer from the Rockefeller Foundation that it would be a "scientific tragedy" if McClintock became unable to continue her work.

All involved agreed on the value of McClintock's research. The Rockefeller Foundation actually made the grant of $1,800 per year to Rollins Emerson, but it indicated that the money was for McClintock to continue her work in his laboratory. The grant was renewed in the following year, but by 1935 McClintock found herself looking for a position once more.

Her former professors and colleagues helped her to find something permanent. It was not an easy task. She was as proud as she was independent. Some other members of the scientific community thought she had "a chip on her shoulder." One problem she had was that she did not want to teach, for two reasons: one, she was not a great teacher, and two, she thought teaching did not give her enough time to conduct research. Really, she wanted only to do her research.

The beginning of 1936 was a time during which McClintock worried. She was less than productive than usual—she would publish no papers at all that year. In the spring, however, matters improved. One of her former colleagues, Lewis Stadler, helped her obtain a position as an assistant professor at the University of Missouri, where he was already on the faculty. The Rockefeller Foundation had given Stadler an $80,000 grant to establish a major center for genetic research at the University of Missouri. He wanted to work on the ring chromosome research and thus was eager to have McClintock as a colleague. He persuaded the University to offer McClintock her first full faculty position. The job offered her a better salary, her own laboratory, and the time and facilities to do her research. With Stadler's support, McClintock was able to continue her work on broken chromosomes.

THE BREAKAGE-FUSION-BRIDGE CYCLE

As soon as McClintock arrived at the University of Missouri, she set to work. She brought her brand of offbeat dedication with her. One episode on the University of Missouri campus made her famous there: She forgot her keys on a Sunday afternoon when the laboratories were locked. Instead of wasting time going back to get them, she simply hoisted herself into the building through a window. A passerby snapped a photograph of the boyish woman in trousers halfway through the open window. Barbara McClintock was never known for her conventional behavior, but she was always known for her superior work and surprising results.

Spending hour after hour bent over her microscope, looking at maize cells, McClintock began to see within the chromosomes several more clues to the mystery of life. Among these was what she called "the breakage-fusion-bridge cycle." *Centromeres* are a specialized region that appear as a thickening or "bulging waist" in the length of a chromosome; during meiosis the centromeres line up at the middle of the cell and move along spindles to either side of the cell. Normal pairs of chromosomes, "sister" chromosomes created during the second phase of meiosis, separate from each other, and move to opposite sides of the dividing cell—preparing to become the centers of

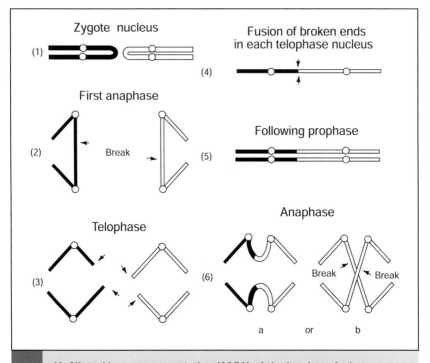

McClintock's own representation (1951) of the breakage-fusion-bridge cycle in plants whose chromosomes have been broken by X-rays. In this cycle, chromosomes with two centers—dicentric chromosomes—are pulled apart, stretch, break, and recombine. The 1938 discovery was of great importance and reinvigorated McClintock's work, which had suffered from her lack of job prospects in the late 1930s. She wrote to a colleague in 1940: "Have been working . . . on an exciting over-all problem in genetics with wonderful results. It gets me up early and puts me to bed late!" (NLM)

the two new cells. McClintock noticed that in one nucleus on one of her slides two broken ends of a chromosome had fused together during meiosis. Each of these broken pieces contained a centromere—so whereas most chromosomes contain only one centromere, this chromosome had had two. While the normal single-centromere chromosomes moved as usual toward their opposite ends of the nucleus, the chromosome with two centromeres had been pulled in *both* directions—

photographs show the chromosome stretched all the way from one end to the other. McClintock called this phenomenon a *bridge*. (Comfort, 73) At the end of this division, the nucleus had pinched itself down in the middle, as nuclei do in this phase of cell division, to complete the actual physical separation into two cells. This pinching had broken the chromosome in two—a phenomenon that McClintock called a *breakage*.

Each of these daughter chromosomes would then, in the next meiotic cycle, undergo replication into a pair of chromosomes. McClintock described the next step: "What happens next is extraordinary and proved to be highly significant for later studies: the two ruptured ends find each other and 'fuse' (are permanently ligated together)." (Fedoroff and Botstein, 205) The broken ends would immediately fuse with each other, thus creating another chromosome with two centromeres, and in future divisions the same cycle of stretching, breakage, and fusing would happen again and again.

McClintock continued to study these special chromosomes for a long stretch of time, into 1939. How, she wondered, did the double-centromere chromosome form? No one knew exactly when in the process of meiosis the chromosomal material was duplicated. McClintock realized that the formation of the chromosome with two centromeres could happen only after the chromosomal material doubled. Thus, duplication of the chromosomes had to have occurred by that time. She began to argue for that conclusion, and she was later proved correct.

Why and for how many generations of cells would this special "breakage-fusion-bridge" (BFB, sometimes *bfb*) behavior continue? Before she was able to answer these questions satisfactorily, she discovered another anomaly—the "breakage-fusion-bridge" continued in the germ cells of the plant as long as meiosis continued but stopped in the cells of the embryo and the resulting maize plant. In the plant, the BFB problem healed itself! Again the questions jumped out at McClintock. She pondered why the BFB ended and how the plant stopped it. It would take McClintock decades to solve these riddles; she would continue to study the ramifications of the breakage-fusion-bridge cycle for another 20 years.

But her colleagues immediately adopted scientific techniques she had developed while working on the problem. By using the breakage-fusion-bridge, McClintock could produce mutations along the length of this one specific chromosome almost wherever she wanted. It was no longer necessary for her to use the clumsy approach of X-rays to generate mutations. Regardless of the results of her further experiments, McClintock, by publishing her techniques and results for others to judge, had given to the genetic community a fine tool that anyone could use to generate site-specific mutations in maize and, by implication, in any other organism. Sometimes, the most important contributions in science are the techniques that are developed, not the actual results of the experiment. Watson and Crick received the Nobel Prize for the brilliant insights they had about the nature and structure of the DNA double helix, but Herman Muller received his Nobel Prize for his techniques of using X-rays to induce mutations.

TELOMERES

McClintock's research raised questions: If broken chromosomes were so ready to combine with any available chromosomal fragments, even to form rings with themselves, why did regular, whole chromosomes not exhibit this reactivity? Why and how did the BFB chromosomes heal themselves once they reached the embryonic maize plant? McClintock hypothesized that there must be something special about the ends of normal chromosomes that prevented them from forming rings or adding other fragments to themselves—some region of inactivity that prevented them from reacting with other chromosomal fragments. In fact, it was *telomeres*, which might be compared to the plastic caps on the ends of shoelaces, that kept the normally highly reactive chromosomal material from reacting. McClintock presented her conclusions in a famous article in *Genetics* in 1941 entitled simply "The Stability of Broken Chromosomes in *Zea Mays.*"

The answers to how and why these broken chromosomes could heal themselves in the embryo of maize were simply beyond even McClintock's ability to determine. She was still doing all of her research with a light microscope; only because she had both incredible eyesight and the courage to believe in what she could only dimly perceive was McClintock able to make the leaps in knowledge that she did. Scientists later discovered that this healing is due to the production of new

A BRIEF HISTORY OF THE MICROSCOPE

When Barbara McClintock began her career as a geneticist, researchers were limited to light microscopes, which had existed in various forms since the end of the 17th century. These microscopes had an important limitation: they could not magnify objects beyond a factor of 500 or 1000. To get a very detailed view of the interior structures of organic cells, scientists needed to be able to magnify their specimens by a factor of up to 10,000. The solution to this was the electron microscope, a kind of microscope that uses a focused beam of electrons instead of light to "see through" the specimen. Ernst Ruska of Germany developed the first of these in 1933; this was the transmission electron microscope (TEM), which worked rather as a slide projector. In 1942 was developed another sort of electron microscope, the scanning electron microscope (SEM), but this was not available for sale until 1965. The Carnegie Institution of Washington donated McClintock's favorite microscope to the Smithsonian Institution after her death in 1992.

McClintock (left) with Harriet Creighton, whom she took under her wing at Cornell. The two women collaborated on such projects as researching how genes can "cross over" in maize to exchange genetic information, and they remained close friends even after their work together. This photograph was taken in 1956.

chromosomal ends, telomeres, and that the enzyme that directs the production of these telomeres, telomerase, is present in the embryo of the plant but absent during meiosis.

Telomeres have become justly famous; they are known to play a much more important role in the life cycle of cells than simply as a cap for the highly reactive chromosomal material. A normal chromosome in the beginning of its life contains a telomere of some great length on each end and the ability to make more. It soon loses the ability to make more. Every time the cell undergoes reproduction, a little bit of the telomere on each end of the chromosome is lost. After a certain number of cell divisions, the cell loses so much of the telomeres that the chromosomes are likely to undergo the destructive changes that McClintock noticed. In this way, the cell loses the ability to make copies of itself that the body can use. Scientists today see the destruction of telomeres as one of the reasons why organisms age and finally die. It is as if organisms have a programmed clock inside each chromosome—and when the clock's cycle reaches its end, the chromosomes self-destruct and the organism dies.

In an amazing example of serendipity, cancer researchers realized at about the same time that cancers have the ability to turn off the telomere clock. A part of the destructive nature of cancer is to make a cell reproduce an infinite number of times; this is the nature of a tumor. If cancer researchers could find a way to stop the infinite reproduction of a malignant cell, they could halt the growth of a tumor. Cancer researchers are now looking for a way to restore the workings of telomeres, while geriatric researchers, those scientists who are looking for a way to prolong human life, are looking for a way to turn telomeres off. All of this was beyond McClintock's view, but her mind reasoned that such a device as a telomere must exist. She lacked the right tools but not the imagination.

Despite this incredibly powerful research, McClintock was not appreciated at Missouri. They remembered too well her eccentricities and too poorly her exciting new work. When McClintock confronted the dean asking about her future at the University of Missouri, he made it clear that she was there only because of Lewis Stadler and that: if anything happened to him she would probably be fired. McClintock requested an unpaid leave of absence and left Missouri in June of 1941, intending never to return.

5

Free to Do Research: 1941–1967

If I could explain it to the average person, I wouldn't have been worth the Nobel Prize.
—Richard P. Feynman, Nobel laureate in Physics (1965), 1985

The kindness and loyalty of friends saved McClintock yet again. She wrote to her old friend Marcus Rhoades, who had just moved to Manhattan and accepted a position at Columbia University. Manhattan is known for many things, but not as a location to grow corn. When McClintock asked Rhoades where he planned to grow his corn, he told her that he would spend the summer on Long Island at Cold Spring Harbor, about an hour away from Columbia's campus. Another former colleague, Milislav Demerec, had been at Cold Spring Harbor for nearly 20 years, and he also held McClintock's work in high regard. Together Demerec and Rhoades arranged for McClintock to be invited to

It was at this time in her life that McClintock realized that the isolation of places like Cornell and Cold Spring Harbor suited her; what she wanted out of life was to explore the mysteries of the maize plant. The work took a great deal of patience—as each new crop, of course, took a year to produce—and McClintock continued in this vein for some 50 years. Her care of her maize was meticulous; she and others often had to put paper bags over the plants to keep the maize from cross-pollinating. This is an image of McClintock's former maize fields as they are labeled today.

Cold Spring Harbor, where she spent the summer of 1941.

Cold Spring Harbor is a small research institution on Long Island, 35 miles from Manhattan, on a secluded inlet off of the Long Island Sound. Founded at the end of the 19th century, Cold Spring Harbor has been a haven for some of the most brilliant researchers in the biological sciences. Much like Ithaca, the remote parts of Long Island have a scenic beauty that seems to enhance the intellectual endeavors that take place there.

Summer came and went, and McClintock stayed on at Cold Spring Harbor until November, when the summer living quarters were closed for the winter. With nowhere else to go, McClintock stayed in a spare room in Marcus Rhoades' apartment in Manhattan.

In early December of 1941, just days before the attack on Pearl Harbor, Milislav Demerec was named director of the Department of Genetics of the Carnegie Institution of Washington at Cold Spring Harbor. Almost immediately, Demerec offered McClintock a one-year position there. She accepted the position, and within a few months Demerec offered to make the position permanent. McClintock had a successful meeting with the president of the Carnegie Institution in Washington, D.C., and he enthusiastically supported offering her a secure, permanent place at Cold Spring Harbor. The Carnegie Institution gave her the money to work there.

It took McClintock a while to realize that Cold Spring Harbor was an ideal place for her. She was free there to do her own research. She didn't have to teach or deal with academic politics or administrative responsibilities. She had a laboratory, a salary, a home, and a place to grow her maize. She was one of only about six or eight full-time, year-round investigators; there were also a few fellows and research assistants. In the summers, the population at Cold Spring Harbor swelled to three times its standard complement of full-time investigators, along with numerous assistants and guests; McClintock later recalled, "It was about four or five years before I really knew I was going to stay." (Keller, 109)

McClintock was lucky to be at Cold Spring Harbor. The reality of World War II hit the United States, and at many other institutions across the nation scientists were forced to abandon the research that most interested them personally to focus on wartime projects. The Carnegie Institution realized the

McClintock in the 1940s with L.C. Dunn, the heir to T.H. Morgan's "fly lab" at Columbia. Dunn, known primarily for his work with poultry and mice, was the author of *Principles of Genetics*, the dominant genetics textbook of the time. He was also an outspoken opponent of the rampant genetics-based racial prejudice of the 1920s and 1930s. With a colleague from Columbia, he co-authored the monumental *Heredity, Race, and Society*—a discussion of the race problem in the United States—in 1946. Dunn later praised one of McClintock's articles as "mark[ing] the highest point attained up until that time in unifying cytological and genetic methods into a single clearly marked field."

long-term value of McClintock's work and enabled her to continue with it.

McClintock felt the effects of the war in other ways. In wartime, the Cold Spring Harbor Laboratory grew even more remote and quiet than before. A gasoline shortage and food rationing affected life there just as it did in the rest of the nation. There were fewer summer visitors in 1942. The annual scientific symposium had a shortened program that summer and was canceled for the next three years. It was a 30-minute walk to the village of Cold Spring Harbor, or a three-mile walk to the nearest store or movie theater in Huntington. There was nothing to do but work.

McClintock worked steadily and productively. She had a major paper on maize genetics published in the journal *Genetics*. Milislav Demerec highlighted McClintock's accomplishment in his annual reports of the Department of Genetics. Her results were published in the annual reports of the Carnegie Institution. Despite her success, McClintock began to feel restless, eager to spend time somewhere other than in her laboratory at Cold Spring Harbor. When an old friend, George Beadle, invited her to visit him at Stanford in 1944, she accepted gladly.

STANFORD UNIVERSITY

George Beadle was McClintock's friend while she was at Cornell. He was a brilliant scientist in his own right. After earning his doctorate in genetics from Cornell University in 1931, Beadle went to work in the laboratory of T.H. Morgan at Cal Tech, where he studied the fruit fly. In 1935, with Boris

THE FUNDING OF SCIENTIFIC RESEARCH IN THE UNITED STATES

Today, as in Barbara McClintock's day, it can cost a great deal of money for a scientist to undertake a research project. Even those who are professors at large universities frequently need funding—their budgets simply will not cover expenses for equipment or assistants' pay. Generally, they must look for grants to pay for their research. Grants sometimes come from the government but seem more frequently to come from private foundations. Barbara McClintock's research, after the years she spent at universities, was paid for by the Carnegie Institution and grants she received, including the annual cash award from the MacArthur Foundation.

Ephrussi at the Institut de Biologie Physico-Chimique in Paris, Beadle designed a complex technique to determine the chemical nature of gene expression in fruit flies. Their results indicated that something as apparently simple as eye color is the product of a long series of chemical reactions and that although genes somehow affect these reactions, they are not the immediate cause of them. After a year at Harvard University, Beadle pursued gene action in detail at Stanford University in 1937.

At Stanford, Beadle worked on a red mold that grew on bread—*Neurospora*. He was hindered in his research by the fact he could not identify its chromosomes because they were extremely small. This type of identification was McClintock's specialty. When she came to Stanford to see Beadle, McClintock worked on the problem for a few days without getting anywhere. She was completely stuck, so she decided to walk to someplace where she could sit and think. She found a bench under some eucalyptus trees on the Stanford campus and sat for half an hour, clearing her mind. When McClintock returned to the lab, she made the breakthrough she had hoped for: she was able to view the seven individual pairs of *Neurospora* chromosomes. Moreover, she was able to see patterns of bands and other details on the chromosomes themselves that enabled her to track the path of the chromosomes through the cycle of cell division. Beadle would claim later, "Barbara, in two months at Stanford, did more to clean up the cytology of *Neurospora* than all other cytological geneticists had done in all previous time on all forms of mold." (Keller, 114)

Working with Edward Tatum, Beadle found that the total environment of *Neurospora* could be varied in such a way that the researchers could locate and identify genetic changes, or mutations, with comparative ease. They exposed the mold to X-rays. Beadle then observed that the mutant molds lost the ability to make a particular organic compound

A 1958 photograph of McClintock's geneticist colleague George Wells Beadle. They had first met at Cornell, and he would later invite her to visit him at Stanford University, where he was studying genes in the bread mold *Neurospora*. Unable to locate the chromosomes himself, he enlisted McClintock, whose skill with microscopes enabled her to identify the chromosomes in two months.

that they needed to digest certain types of food. The molds would starve to death. Beadle determined that the function of each gene was to control the production of a particular enzyme that the organism would then use in some bodily function. This "one gene–one enzyme" concept won Beadle and Tatum the Nobel Prize in 1958.

McClintock's work for Beadle had other repercussions as well. No one had been able before to view the entire reproductive cycle of any fungus. It is amazing that McClintock was able to do this given the technological limitations of the time. She had to prepare separate slides over and over again, as the equipment did not exist that would allow her to view the action in real time.

One of the things that made McClintock so much more successful than other scientists was her ability to lose herself in her work. Ever since childhood, she had shown an intense focus and concentration on whatever interested her, to the exclusion of everything and everyone else around her. McClintock later explained, "As you look at these things, they become part of you. And you forget yourself. The main thing about it is you forget yourself." (Keller, 117)

The year 1944 proved very good for McClintock. Her success at Stanford with the mold *Neurospora* was only one of the highlights. At 42 years old, she was finally achieving the professional recognition she longed for. Her election to the National Academy of Sciences was a public acknowledgment of her accomplishments. She was only the third woman elected to the Academy. The same year she was elected president of the Genetics Society of America. She was the first woman to hold that office. When McClintock returned to Cold Spring Harbor that winter, she was in good spirits and filled with confidence, and it was during this time that she began the most important and controversial work of her career.

McClintock returned to Long Island and Cold Spring Harbor late in 1944. Waiting for her was the maize that had grown that summer, and now she started to investigate something new. McClintock began to experiment with crossing corn from two different parents, each of which contained a different BFB chromosome 9. She planted 677 kernels of such a cross in the summer of 1944. She noted the physical characteristics of these mature plants and then self-pollinated those that survived the summer. She examined this second-generation crop of kernels for mutations. She planted 45 mutant kernels indoors early in 1945 and allowed their seedlings to grow to maturity. The leaves of these plants showed a whole range of patterns and differences. It was a totally unexpected result. She noticed in this crop that there were variations—mutations—from the normal green color. Some seedlings contained discolored patches of white, some of light green, and some of pale yellow. These patches showed genetic instability, or *mutable genes*, because the parents of these discolored seedlings were solid in color and a genetic mutation had occurred to cause the seedlings to appear different. McClintock commented:

> I soon recognized that the changes in patterns of variegation that appeared in sectors on these new leaves held the key to an understanding of the events that were responsible for initiating variegation in the first place. Most significant in this regard were twin sectors, obviously derived from sister cells, in which the pattern changes in the twins were reciprocals of each other. For example, a reduced frequency of mutations to give a full chlorophyll expressions on the pale or white background in the surrounding leaf tissue was matched in the twin with a much increased frequency of such mutations. My conclusion from these twin sectors was that during a mitotic cycle one cell had gained some component that the sister cell had lost, and that this component was responsible for regulating (i.e., controlling) the mutation process: that is, its time and its frequency of occurrence in plant tissue. (Fedoroff and Botstein, 207)

To put it much more simply, genes from one chromosome had moved to another chromosome. The big discovery for McClintock was that these mutations did not occur randomly but were regulated by some other substance within the cell nucleus. There was a pattern. By observing a mutation such

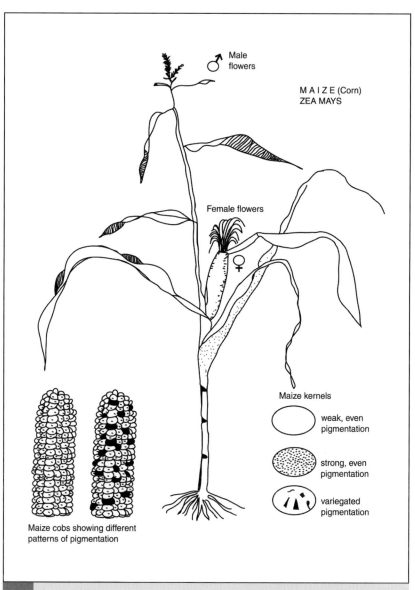

Male flowers

M A I Z E (Corn)
ZEA MAYS

Female flowers

Maize kernels

weak, even pigmentation

strong, even pigmentation

variegated pigmentation

Maize cobs showing different patterns of pigmentation

This drawing of a maize plant from the 1983 Nobel press release also shows variations in cobs and kernels. Corn plants are self-pollinating and therefore have both female and male reproductive parts. The tassel at the top of the plant is composed of anthers, which scatter pollen onto the female flowers (or silk) on the tops of the cobs below.

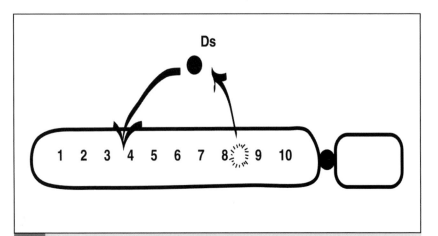

In this diagram, a controlling element in a chromosome "jumps" from one position to a new position between different genes. This is how genes are "switched on and off"; a gene that has been "switched off" will stop directing the production (synthesis) of the protein in controls. Sometimes this causes instability in the chromosome, too, causing it to break more easily. McClintock was surprised to find that this "jumping" of controlling elements was not as random as it might seem.

as a color difference, McClintock could trace the genetic history of the plant. Since the mutations occurred regularly, there must be something controlling the rate of mutation. No one had yet seen a gene, there was no proof of their existence— so for McClintock to hypothesize that there was some substance within the cell nucleus that regulated the *expression* of the gene was fantasy. No one would believe her. In fact, no one did.

She named the controlling element that caused the chromosome carrying the genes for characteristic color to break "Dissociator." Her experiments showed that the chromosomal breaks always happened close to Dissociator but that the location of Dissociator changed and it caused breakage at different locations along the chromosome. She found that Dissociator moved only if another controlling element, which McClintock called "Activator," also was present. She futher found that Activator moved around in the cells of the organism.

McClintock believed that controlling elements explained how complex organisms could develop many different kinds of cells and tissues when each cell in the organism had the same set of genes. The answer lay in the regulation of those genes. For six years, McClintock recorded data to support her observations. Her office was filled with stacks of notes and data to answer every objection she could anticipate from her fellow scientists. Between 1948 and 1950, McClintock developed a theory: that these transposable elements regulated the genes by selectively inhibiting or modulating their action. Finally, in 1950, McClintock published her results.

She had been publishing some of her work in the Carnegie yearbooks, although there were other scientific journals that probably would have been eager to publish her papers. She chose to publish the major results of her research on mutable genes in *Proceedings of the National Academy of Sciences* in June of 1950. McClintock had so much data that it could not all be included in the published article. This was one of her first published articles that focused on theory and interpretation, rather than on data and evidence. Her confidence in her theories was unshakable.

McClintock's first public presentation of transposable elements—sometimes referred to as "jumping genes," especially in the popular press—was in the summer of 1951 at the annual Cold Spring Harbor Symposium. The topic of the conference was "Genes and Mutations." With two decades of pioneering cytogenetics to her credit, McClintock now had a strong scientific reputation at Cold Spring Harbor. Even at such a renowned institute, McClintock stood apart from the crowd of nearly 300 researchers. First, she was one of the only geneticists who still worked on corn. Studying maize and *Drosophila* was no longer routine, for old organisms had been replaced by studies of bacteria and viruses that reproduced faster.

McClintock had also fashioned for herself a unique persona. She was already in the minority as a woman, but she also had "an exotic flamboyance as she lounged on the patio between sessions, dressed in her white shirt and khaki slacks, smoking cigarettes with a long holder." (Comfort, 157) (See photo, page 75.) Her sharp wit and great intellect intimidated some, but those who were bright enough to engage her in conversation were rewarded with the passion and intensity of her ideas.

CHALLENGING THE ACCEPTED THEORY

By 1952, R.A. Brink at the University of Wisconsin and P.A. Peterson, later at Iowa State University, had each published articles in the scholarly journals that confirmed the existence of transposable elements in maize. Still, there were plenty of geneticists who doubted the validity of the theory. McClintock reflected later,

> In retrospect, it appears that the difficulties in presenting the evidence and arguments for transposable elements in eukaryotic organisms were attributable to conflicts with accepted genetic concepts. That genetic elements could move to new locations in the genome had no precedent and no place in these concepts. The genome was considered to be stable, or at least not subject to this type of instability. A further difficulty in communication stemmed from my emphasis on the regulatory aspects of these elements. In the mid-1940s there was little if any awareness of the need for genes to be regulated during development. Yet it was just this aspect that caught my attention initially. . . . It was not until fifteen years later that the regulation of gene action began to gain credibility due to the elegant experiments of Jacob and Monod that were carried out in bacteria. (Fedoroff and Botstein, 208)

In 1951, these ideas were as unconventional as McClintock herself. While her colleagues were ready to accept the idea of transposition and mutations, they were not so sure about the idea that the same elements also guided the development of the whole plant. Her presentation at the 1951 Cold Spring Harbor Symposium, full of statistics and proofs, lasted more than two hours. When it ended, McClintock later recalled, her lecture was greeted with "puzzlement, even hostility" from her audience. She felt that "nobody understood." She'd anticipated questions, but there had been few.

Several people who were there disagree with McClintock's recollection that her work was not appreciated. Nobel Laureate Joshua Lederberg was present. "Between 'stony silence' and 'instant appreciation,'" Lederberg argued later, "is the reality of *how* to integrate the startling evidence she presented into a coherent scheme. That was hardly possible before . . . the science of molecular biology caught up with [McClintock]. Perhaps some of the *biochemists* in the 1950s were not well versed in maize genetics and it is their voices [we] hear." Lederberg also pointed out that the symposium organizers, her boss, Milislav Demerec, for one, must have recognized the importance of her work, or she would not have been invited to speak and given precious time on the very full agenda.

McClintock's theory was later discovered to be correct. Other transposable elements have been found in maize. The decorative corn with hundreds of different colored kernels that is seen around Halloween and Thanksgiving is the result of selection for transposable elements. In fruit flies it was found that there are about 50 controlling elements, now known to be sequences of nucleotides of approximately 5000 base pairs. Transposable elements are responsible for the wrinkled-skinned peas that Gregor Mendel studied. But even today, "the molecular processes responsible for the movement of transposable elements are not well understood." (Hartl, 134–135) Their purpose within the cell is unknown; they remain one of the puzzles of heredity.

McClintock seems to have been hurt by her colleagues'

apparent lack of understanding. Nevertheless, she continued to work undisturbed at Cold Spring Harbor. Perhaps they didn't agree with all her theories, but colleagues invited McClintock to visit and lecture often. Throughout the 1950s and 1960s McClintock was invited to give lectures on her theory of controlling elements, as well as on important general themes

WHY BARBARA MCCLINTOCK STOPPED PUBLISHING

In May of 1973, Dr. McClintock explained in a letter to J.R.S. Fincham of the University of Leeds the reasoning behind her longstanding reluctance to share her maize research with the scientific community:

> . . . I recognize the degree to which many aspects of my reports are not comprehended by many of those working with my materials or with similar or related ones. Much of this is my fault. I stopped publishing detailed reports long ago when I realized, and acutely, the extent of disinterest and lack of confidence in the conclusions I was drawing from the studies. With the literature filled to the exhaustion of all of us, I decided it was useless to add weight to the biologist's wastebasket. Instead, I decided to use the added time to enlarge experiments and thus increase my comprehensions of the basic phenomena. . . .
>
> All of the above is not intended as a complaint. Rather, it is to let you know why I stopped publishing detailed accounts after 1953, and also and particularly because I wish you to know how much I have appreciated your careful considerations and your thoughtful comprehension of the substance of these summaries. Such comprehension has been rare, indeed. . . . (NLM, 2002)

in maize genetics, at universities around the U.S. For example, in the winter of 1954 McClintock was invited to lecture over the course of an entire semester at Cal Tech.

THE ORIGINS OF CORN

Maize was not only the "guinea pig" for all of Barbara McClintock's important scientific discoveries; it is also corn on the cob, one of America's summertime favorites. For over 8,000 years, corn has been a major food of North and South America. By 2000 B.C., corn had been traded to the Indians of present day United States and grown there. The use of corn in the United States had the same effect as it had in Mexico—it allowed a small percentage of the population to feed the rest, freeing them to become artisans, warriors, kings, and politicians. After the discovery of the Aztec civilization around 1520, corn was sent to Europe, where it was grown and used as food. The use of corn for food and the knowledge that corn could be stored in the event of emergency or winter was perhaps the major reason why the great civilizations such as the Olmec, Mayan, and Aztec developed in Mexico. But where, asked scientists, had corn come from? There was no wild corn growing in Mexico; all of the corn that had ever been known was domesticated corn. Humans must have "invented" corn.

At one time, in central Asia 10,000 or 20,000 years ago, there was an edible green plant that looked a lot like kale does today. It had large, wide, edible green leaves that women picked and fed to their families. At the dawn of agriculture, however, women began to save the best plants and to plant their seeds in the following spring. They would do this every year, and in a few generations they had kale plants that were bigger, greener, and more insect-resistant than any that could be found in the countryside. In addition, some saw that some plants had more flowers than others and that the flowers themselves were good to eat, especially while they were still green and tender. They began to choose for this characteristic, too. Soon they had kale plants that had especially tasty leaves and kale plants that had especially abundant and tasty flowers. They stopped calling the latter *kale* and began to call it *broccoli*. Again, they chose kale plants that had tiny flower buds that arose along the stem of the plant, and in a little while they had what came to be known as *Brussels sprouts*. Further selection over time for the leaves resulted in cabbage. Cauliflower was another selective creation of untold and unknown generations of early farmers.

When the American Indians arrived in what became Mexico, probably about 15,000 years ago, they were hunters of mammoths, elk, and other large animals. The women were gatherers; they picked the flowers of wild amaranth and the flat leaves of the prickly pear cactus. They took the fibers of maguey and wove baskets, clothing, and shoes. They also picked a local grass that was found in central Mexico, called *teosinte*, and added its small kernels to soups. These were tasty, but they were tiny and each grass plant had only a few kernels. Over the next 5,000 years or so, the peoples of central Mexico selectively bred teosinte grass into what is now called *corn*.

In the 1930s, the story of corn was still untold. It was a very interesting problem, and some great minds, winners of several Nobel Prizes, turned their attention to its solution. George Beadle and Paul Christof Mangelsdorf proposed two contrasting theories for the origin of corn on the cob. Beadle proposed what came to be called the "teosinte hypothesis," in which corn had been domesticated from teosinte by human selection. He published his theory in the article "Teosinte and the Origin of Maize," published in *Journal of Heredity* in 1939. Mangelsdorf a few years later published a conflicting article, "The Origin and Evolution of Maize," in *Advances in Genetics*. Mangelsdorf's "tripartite hypothesis" was that modern corn was a result of a cross between teosinte and another corn-like grass, *Tripsacum*. To some, the debate was trivial. Mangelsdorf's theory supposed that, one day in the

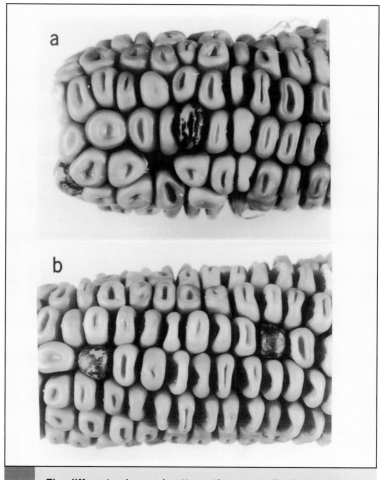

The different colors and patterns that are easily observed in maize—often called "Indian corn"—are what make it an ideal plant for studying genetics, since these colors are dependent on inherited genes. Geneticists later moved on to study fruit flies and then microorganisms, which have shorter life spans, making it easier to study how genes are passed on in a shorter time period.

remote past, an unknowable Mexican Indian had shaken the pollen from *Tripsacum* grass onto the silk threads of teosinte. (Less romantically, Mangelsdorf's cross could have simply been caused by wind.) Beadle's theory celebrated the hundreds of generations of patience and determination of the human spirit. Each side had its proponents. From the 1930s through the 1960s, the majority of opinion favored Mangelsdorf's view.

During World War II, the United States began a program to help the country of Mexico to become self-sufficient in food production. One of the facets of this program was to develop new, more productive strains of corn, the major foodstuff of Mexico. The United States government asked the philanthropic Rockefeller Foundation to provide scientists and resources for such a program. The Rockefeller Foundation began its research into the ancestry of maize as a part of this project and continued it after the end of World War II and well into the 1960s. During her tenure at the Carnegie Institution, McClintock was also a consultant to the agricultural science program of the Rockefeller Foundation, which funded research in maize in Mexico.

THE ROCKEFELLER FOUNDATION

This was not the only time that McClintock's career benefited from the generosity of the Rockefeller Foundation. The industrialist and philanthropist John Davidson Rockefeller started the Rockefeller Foundation in 1913. Rockefeller was born into poverty, the son of a peddler, but went on to make millions as the founder of Standard Oil. He believed, as did his fellow philanthropist Andrew Carnegie, that it was foolish for someone of his wealth to wait until after his death to use his money to achieve good things. The Rockefeller Foundation's mission is to enrich and sustain the lives and livelihoods of poor and excluded people throughout the world. To fulfill it's mission, the Rockefeller Foundation has concentrated on fighting the war against poverty, hunger, and disease throughout the world, often providing education and employment. Other primary concerns of the Foundation are overpopulation, environmental conservation, and support of the cultural and creative arts. The work of scientists and scholars sponsored by the Rockefeller Foundation led to many improvements in public health and food production in the 20th century. Beginning in 1933 and continuing for more than 20 years, the Foundation spent $1.5 million to identify 300 scientists and scholars from Nazi Germany and help them to settle in friendly locations, including many American universities. Since its inception, the Rockefeller Foundation has given more than $2 billion to thousands of grantees worldwide and has assisted directly in the training of nearly 13,000 Rockefeller Foundation Fellows. The Rockefeller Foundation has set up laboratories in La Molina, Peru; Medellin, Columbia; Chapingo, Mexico; and Piricicaba, Brazil. It has established biologists trained in the United States in these laboratories to conduct research into, among other things, the ancestry of maize.

SOUTH AMERICA

In the spring of 1957, Paul Mangelsdorf, who worked closely with the Rockefeller Foundation, was visiting at Cold Spring Harbor. At the end of his visit, McClintock drove him to the nearby train station. She later remembered, "Just as we were getting near the station he said that he would like to have somebody in Peru trained in cytology, and asked me would I be interested. And I said, just as we got to the station, 'Yes.'" McClintock left for Peru on December 5. She spent about six weeks in La Molina and in Lima, Peru. She prepared slides to reveal maize chromosomes, showing the local students how to do this. She then analyzed the prepared slides and showed the students how to identify the chromosomes by their size and shape. She met with Mangelsdorf in February of 1958 and reviewed her work with him. He was so pleased with the work that he invited her to return to South America, this time to the laboratory in Medellin, Columbia.

In December, after the 1958 growing season at Cold Spring Harbor, she flew to Columbia. She began examining corn that

McClintock doing research in Mexico in 1959, around the time she was also studying different types of corn growing in South America. Her studies of South American corn led her to hypothesize about the origins of the plant, a point that had long been a subject of debate.

had been collected in Ecuador, Bolivia, Chile, and Venezuela. Quite soon after she began to examine the corn under the microscope, McClintock was struck by one amazing fact: Throughout the entire geographic region from Chile to Columbia, almost all of the corn that had come from high altitudes was of the same race—9 of 10 high-altitude samples from Ecuador, 11 of the 12 from Bolivia, and all 10 from Chile. From the corn samples that were grown in lowland areas throughout the same region, many different races of corn were found. She hypothesized that it would be possible to trace the origin and distribution of corn through an examination of the chromosomal changes found in corn samples from throughout its growing range. She also thought that her work might lead to commercial successes in corn breeding. McClintock and others continued the South American research well into the 1970s. In 1981, with Rockefeller Fellows Almeiro Blumenschein and Angel Kato, McClintock co-authored *The Chromosomal Constitution of Races of Maize.*

Although McClintock seems to have found happiness in her work in Latin America, her letters of the time to colleagues such as Curt Stern and George Beadle focus almost entirely on her interest in the races of corn. She gives no clue to how she fared as an older woman living and working in harsh, undeveloped locations. The easy conclusion is that at this time, just as in her earlier years, McClintock was more interested in her research than in the creature comforts of life.

In 1967, at the age of 65, McClintock received the Distinguished Service Award from the Carnegie Institution of Washington, Department of Genetics, Cold Spring Harbor. She was officially retired, but she retained the title of scientist emerita and continued to work at the laboratory at Cold Spring Harbor. In fact, she would continue to work until four months before her death—6 days a week and up to 14 hours a day. Cold Spring Harbor provided her with all she needed— a home, a laboratory, and, most important, recognition of the importance of her work.

6

Recognition at Last: 1967–1983

I never thought of stopping, and I just hated sleeping. I can't imagine having a better life.
—Barbara McClintock, 1983

McClintock was gaining positive recognition from the scientific community, which realized the great significance of her work. The Distinguished Service Award that McClintock received from the Carnegie Institution was not the only recognition she would receive in 1967. She also received the Kimber Genetics Award from the National Academy of Sciences in that year—the highest honor that could be given to a geneticist. McClintock's win was particularly prestigious because the Kimber was awarded by a committee of prominent geneticists, her colleagues.

Just two years earlier, McClintock's alma mater, Cornell University, had appointed her to the honorary position of

After winning the Nobel Prize in Medicine or Physiology in 1983, McClintock found herself in the public eye, a position she did not relish. Although she did speak at press conferences and give the expected interviews, she much preferred being in the lab and doing her research.

Andrew White Professor-at-Large. Other universities acknowledged McClintock's scientific achievements over the years, too. She was awarded honorary Doctor of Science degrees by the University of Rochester in 1947, Western College in 1949,

Smith College in 1957, the University of Missouri in 1968, Williams College in 1972, and The Rockefeller University and Harvard University in 1979. Georgetown University presented her with the degree of Honorary Doctor of Humane Letters in 1981. Her achievements were also recognized by the scientific community as a whole—as in 1971, when then President Richard M. Nixon honored her with the National Medal of Science, the highest science award the American government gives.

Molecular biology is the study of the molecules that direct molecular processes in the cell. It focuses on the physical and chemical organization of living matter and especially the molecular basis of inheritance and the creation of proteins. The American mathematician Warren Weaver first used the term *molecular biology* in 1938 to refer to "those borderline areas in which physics and chemistry merge with biology." Using sophisticated tools such as electron microscopes, molecular biologists became able to see clearly the phenomena that McClintock had seen with her old-fashioned microscope and slides.

In 1973 McClintock explained in a letter to fellow maize geneticist Oliver Nelson, "Over the years I have found that it is difficult if not impossible to bring to consciousness [in] another person the nature of his tacit assumptions. . . . One must await the right time for conceptual change." (NLM, 2002) Conceptual changes were happening partly because molecular biology made it possible at last for others to see what McClintock and only a handful of her peers had understood more than 30 years earlier.

The world had begun to realize the complexity of the study of genetics only in 1953. That year, biologists James D. Watson of the United States and Francis H.C. Crick of Britain proposed a model of the double-helix structure of deoxyribonucleic acid (DNA). Genes, such as the ones that McClintock studied, are composed of DNA, so DNA plays a central role in cellular division. The importance of Watson and Crick's work was recognized quickly. In 1962, they and their colleague Maurice Hugh Frederick Wilkins shared the Nobel Prize in Physiology or Medicine.

In 1981, McClintock became the first person ever to receive a MacArthur Foundation grant. Every recipient of the MacArthur award, which has been referred to as the "genius grant," receives an annual fellowship for life of $60,000, tax-free. The MacArthur Foundation's explanation of the grant speaks to McClintock's merit:

> The MacArthur Fellows Program awards unrestricted fellowships to talented individuals who have shown extraordinary originality and dedication in their creative pursuits and a marked capacity for self-direction. There are three criteria for selection of Fellows: exceptional creativity, promise for important future advances based on a track record of significant accomplishment, and potential for the fellowship to facilitate subsequent creative work.

Much like employment at Cold Spring Harbor, the MacArthur grant came without restriction. McClintock was allowed to use the money as she saw fit. In that same year, McClintock also received $50,000 from Israel's Wolf Foundation and the Albert Lasker Award for Basic Medical Research. The Lasker Award was especially exciting because winners often went on to win the Nobel Prize.

THE NOBEL PRIZE

The Nobel Prize has been awarded every year since 1901. It is named after the Swedish scientist and businessman Alfred Nobel, who lived from 1833 to 1896. Nobel was an intellectual man who was a master of both science and literature. He held more than 350 patents and composed both poetry and drama. In 1866, Alfred Nobel invented the product that would make his fortune—dynamite. He patented his invention

Dr. Joshua Lederberg in 1958, the year he won a Nobel Prize for his work on the organization and recombination of genes in bacteria. Lederberg, who wrote McClintock's nomination letter, has argued for a different interpretation of the failure at the 1951 symposium at Cold Spring Harbor—that no one was hostile to McClintock's work, and many saw its importance, but that those in attendance needed time to understand it.

and developed it into companies and laboratories in over 20 nations on 5 continents. Dynamite was used in developing nations for mining, road building, and tunnel blasting. Perhaps because he profited so hugely from the invention of dynamite, a substance that can be used for destruction, Nobel was also dedicated to the promotion of peace. In his will, Nobel established the Nobel Foundation. His goal was to recognize those individuals who made significant contributions to five fields—peace, chemistry, literature, medicine or physiology, and physics. A Nobel award for economics was established later, in 1968. The Nobel Prize consists of a medal, a personal diploma, and a monetary award and is presented each year in Stockholm, Sweden.

Although the Nobel Prize is a rare and important honor, McClintock knew several of her close collagues had been honored over the years. Her old friend from Cornell George Beadle had won in 1958 with Edwin Tatum for their discovery that genes act by regulating definite chemical events (the "one gene–one enzyme" theory discussed earlier).

Beadle and Tatum had shared the Prize that year with another of McClintock's acquaintances, Joshua Lederberg. Lederberg had won for his discoveries concerning genetic recombination and the organization of the genetic material of bacteria. James Watson, who won the shared Prize in 1962 for discovering the helical structure of DNA, was McClintock's supervisor at Cold Spring Harbor in the 1970s.

In 1981, some of these same friends wanted McClintock to be nominated for the Nobel Prize. Beadle and Marcus Rhoades, who had supported McClintock for so many years, both wanted to see her win. Beadle, once skeptical of her work, felt the recognition was overdue. Because the nomination had to come from a Nobel laureate, they asked James Watson to make it; but he declined because he had already nominated someone else. Instead, Nobel laureate Lederberg wrote McClintock's nomination letter to the Nobel Committee.

Other nominations of McClintock were also submitted. Some suggested that she should win a lifetime achievement award. Others thought that she should win for her important discovery of transposable elements. The discussions of the Nobel committee are secret, and McClintock's records will be sealed until 2033, but letters from scientists of the time show that McClintock's nomination generated much discussion and many strong opinions about the merits of her work.

On October 11, 1983, the day after McClintock's Nobel Prize was announced, the president of the Carnegie Institution of Washington, James D. Ebert, wrote to her. "Never has a Prize been more richly deserved," he wrote. "And, never has anyone better exemplified the reasons for the [Carnegie] Institution's existence—to seek out and support the uncommonly creative individual who is engaged, in Mr. Carnegie's own words, in 'basic research of a pioneering nature.'"

"THE KATHARINE HEPBURN OF SCIENCE"

McClintock, now 81 years old, didn't enjoy being the object of all this attention. She didn't really need the money; her life at Cold Spring Harbor was very simple, and her needs were very few. She didn't like publicity or crowds. She wanted what she always prized most—her freedom. She wished to be able to eat and sleep and do her research when she chose and how she chose. The only thing that really gave her joy was solving the mysteries of nature.

Unfortunately for McClintock, the media loved her. She was a little old lady who looked like a tomboy. James Watson had dubbed her "the Katharine Hepburn of science" a few years earlier, and the title seemed very fitting now. Everyone wanted to interview her. She was outspoken and had always known that she was right, even when no one was giving her prizes and awards; she would have been happier if everyone had just understood the meaning and value of her work many years earlier. Through it all, though, McClintock kept her sense of

Although there was reluctance to accept McClintock's work early in her career, it seems clear that most knew she had great potential from the start. When full recognition came at last, it came in a flood and actually embarrassed the naturally reticent McClintock. It was not unusual for her to avoid the ceremony that accompanied honors; fortunately, her position at Cold Spring Harbor eliminated the concern about funding for her research.

MAIZE RESEARCH AFTER
THE "GOLDEN AGE" AT CORNELL

A second period in corn studies, inaugurated in the late 1960s, has researchers focusing on describing the diversity of and evolutionary relationships among the species in the genus *Zea*, determining the genes involved in domestication and the mechanisms by which the genome evolves. Garrison Wilkes carefully described the Central American teosinte (*Zea mexicana*) in a monograph published in 1967. Hugh Iltis, John Doebley, Raphael Guzman, and B. Pazy invigorated teosinte research by discovering and describing the perennial species *Zea diploperennis*. Iltis and Doebley established an organized taxonomy, a kind of corn "family tree," in 1980. In 1981, McClintock and some colleagues organized and published data on the diversity of chromosomal "knobs"—the variations in the surface structure of chromosomes that had enabled McClintock to identify the ten chromosomes of maize some 50 years earlier—that had been collected over the previous 30 years. Throughout the 1980s and 1990s, Charles Stuber and Major Goodman's research groups produced comprehensive analyses of the diversity in over 1,000 kinds of maize and in almost all known teosinte populations. Since the advent of DNA analysis and sequencing in the 1980s, John Doebley, Ed Buckler, and Brandon Gaut have refined the understanding of the relationships among grasses and with *Zea*, and they have contributed to the knowledge of how the genome has evolved. In the 1990s, John Doebley's research group began to discover some of the genes involved in maize domestication. (Modified from Buckler, 2002) humor. A photograph taken at Cold Spring Harbor shortly after the Nobel announcement shows McClintock's humorous attempt to go unrecognized: she's wearing a plastic Groucho Marx disguise—eyeglasses, nose, and mustache.

McClintock had never lost her strong belief in herself and her work. In 1983 she described the pleasure of conducting research on her own terms: "Over the many years, I truly enjoyed not being required to defend my interpretations. I could just work with the greatest of pleasure. I never felt the need nor the desire to defend my views. If I turned out to be wrong, I just forgot that I ever held such a view. It didn't matter." (NLM)

Such confidence must certainly have been an asset to one who spent her life exploring the microscopic world and understanding it in ways that few of her colleagues could follow.

Barbara McClintock's Legacy

We believe that Dr. McClintock is without a doubt the most
outstanding cytogeneticist of this country and is hardly
surpassed by anyone elsewhere.
—David R. Goddard and Curt Stern, 1944 (NLM)

Throughout McClintock's life, her friends and supporters helped
to make the difference between her success and failure.
Through moral support, intellectual exchanges, job offers, and
award nominations, McClintock's friends and colleagues
showed their fondness and respect for McClintock. Though she
was often alone, she seems rarely to have been lonely. She often
exchanged letters with her friends and peers that show her
warmer, more sensitive side. In her old age, she especially
enjoyed meeting the young scientists who came to Cold Spring
Harbor especially to see her, to ask her opinion of an idea or
question her about a scientific technique she had developed.

McClintock in 1990, not long before her death in 1992. Looking back on nine decades, the geneticist concluded with obvious satisfaction that she had lived a very interesting and fulfilling life. Among her many achievements is her happiness itself.

Often they came away having had a conversation with her that covered not just science, but also philosophy, art, or politics, all of which interested her almost as much. Children who lived in Cold Spring Harbor would later remember that from time to time she would fall in with them when they went on a walk. She liked to share with them her curiosity about the natural world.

When Barbara McClintock turned 90, there was a party held in her honor at Cold Spring Harbor. People came from seemingly everywhere to celebrate with her. By that time, she had already been the subject of a full-length biography. Now another book was to appear in her honor: Nina Fedoroff's *The Dynamic Genome: Barbara McClintock's Ideas in the Century of Genetics.*

Just a short time later, on September 2, 1992, Dr. McClintock died. She had been ill for a short time. She left behind no husband, children, or grandchildren; her legacy was one of hard work, groundbreaking research, and good friends. A memorial service was held on November 17, 1992, at Cold Spring Harbor Laboratory to celebrate McClintock's life. Friends and colleagues remembered her with warm words and stories. James Shapiro, a friend and peer from the University of Chicago, spoke about McClintock's many contributions to the fields of genetics and cellular information processing. Shapiro told the audience that he "believed that the secret of McClintock's success, in the face of incomprehension and prejudice, was her fearless and complete intellectual freedom—to admit 'I don't know,' and then to wrestle the answer from the data." Shapiro viewed McClintock as "a visionary, a bridge to a new era of biological thought." (Comfort, 267)

Maize geneticist Oliver Nelson came from the University of Wisconsin at Madison to pay his respects. Nelson concluded with his own thoughts on McClintock's successful career: "Researchers are occasionally presented with bizarre results. McClintock possessed a special talent to recognize the underlying order and provide an explanation for the most perplexing observations."

Others remembered the warmth of McClintock's personality. Evelyn Witkin of Rutgers University was a bacterial geneticist who worked at Cold Spring Harbor from 1945 to 1955. During that time she became a close friend of McClintock's. Dr. Witkin shared a story that revealed McClintock's sense of humor:

Henry Kissinger, then Secretary of State, held a dinner for Nobel Laureates, including McClintock. The invitation listed all the other guests as "*Dr*. So and So," but McClintock was listed only as "*Ms*. B. McClintock." McClintock borrowed a line from comedian Rodney Dangerfield and wrote in the margin of the program, "I don't get no respect!" (Comfort, 29)

Despite McClintock's belief that graduate students should "sink or swim," the younger scientists remembered her fondly. V. Sundaresan of Cold Spring Harbor was working toward a Ph.D. at the time McClintock won the Nobel Prize. He recalled that she was always approachable. In fact, he noted that "if you asked her a question, you had better be prepared to confer for a whole afternoon. . . . After several hours of intense dialogue, she would look closely at you and say, 'We'd better stop now—you look tired!'" Even in her 80s, McClintock could out-think and out-talk someone young enough to be her grandchild.

Even James Watson, a director of the Cold Spring Harbor Laboratory with whom McClintock sometimes clashed, called McClintock one of the three most important figures in the field of genetics. Coming from one of the discoverers of the structure of DNA, this is no small praise. A Cold Spring Harbor spokeswoman agreed: "Her discovery was thirty years ahead of its time." ("Barbara McClintock")

In the years that have passed since her death, she has not been forgotten. The American Philosophical Society of Philadelphia—one of the oldest scientific institutions in the United States, founded by Benjamin Franklin before the American Revolution—became the repository for her papers, including notebooks she filled with data and letters she exchanged with other scientists. The National Library of Medicine cooperated with the Society and digitized some of her most important papers, in order to post them on the Internet for use by researchers. Cornell University regards McClintock as one of its most distinguished alumni.

McClintock attends the dedication of the McClintock Laboratory, which was renamed in her honor in 1973, at Cold Spring Harbor. It was at this lab that she first developed the theory of transposable elements that regulated genes in chromosomes—the theory that really brought her work to the public's attention. Cold Spring Harbor was home to her, and it was a godsend—the Carnegie Institution of Washington funded her fully to continue her research there in peace, away from the academic politics that had caused her problems in the past. It took a while to settle in, but it was an ideal setting for the life McClintock wanted to lead.

PAVING THE WAY

It is impossible to say where the study of genetics and medicine would be today without the work of Barbara McClintock. The Human Genome Project, begun in 1990, is an international scientific program established to analyze the human genome, the complete chemical instructions that control heredity in human beings. Someday this "gene mapping" will provide the understanding necessary to unlock the mysteries of diseases such as diabetes and Alzheimer's disease.

Other scientific advances in the areas of cancer research, immunology, genetic engineering, and cloning most likely could not have been achieved without the groundwork laid by McClintock. Her work on the breakage-fusion-bridge cycle later became the basis for interpreting radiation sickness in humans who have been heavily exposed to radiation. Every day, people benefit from improvements in their health and wellness that would never have been possible without McClintock's years alone in her fields, studying her corn.

And conducting her research brought her joy: "I've had such a good time, I can't imagine having a better one," she said at a Nobel press conference in 1983. ". . . I've had a very, very satisfactory and interesting life." (Comfort, 269)

Chronology

1866   Gregor Mendel publishes his research on the genetics of pea plants.

1871   Charles Darwin publishes *Descent of Man*, discussing for the first time the role of sexual selection in heredity.

1888   Heinrich Wilhelm Gottfried Waldeyer coins the term *chromosome*.

1898   Thomas Henry McClintock and Sara Handy are married; Marjorie McClintock is born in October.

1900   Birth of Mignon McClintock.

1902   Barbara (originally Eleanor) McClintock is born on June 16 in Hartford, Connecticut.

1903   Birth of Malcolm Rider (Tom) McClintock on December 3.

1908   The McClintock family relocates to Brooklyn.

1918   Barbara Graduates from Erasmus High School; enrolls at Cornell University, Ithaca, New York.

1921   Studies genetics with C.B. Hutchinson, who invites her to join a graduate course in early 1922.

1923   Receives B.S. from Cornell University in June.

1925   Receives M.A. from Cornell University.

1927   Receives Ph.D. in botany from Cornell University; begins term as Instructor in Botany at Cornell.

1929   Publishes paper that "laid the cytological foundation for all later work in corn."

1931 Becomes a fellow of the National Research Council. Proves that certain chromosomal anomalies that the community has long suspected actually do exist. With Harriet Creighton, publishes proof of genetic crossing over.

1933 Becomes a fellow of the Guggenheim Foundation.

1934 Becomes a research associate at Cornell University.

1936 Becomes assistant professor at University of Missouri.

1939 Elected vice president of the Genetics Society of America (founded 1931).

1941 Leaves the University of Missouri.

1942 Joins staff of Department of Genetics, Carnegie Institution of Washington, based at Cold Spring Harbor, New York.

1943 Changes first name officially to Barbara.

1944 Elected third female member of the National Academy of Sciences.

1945 Elected president of the Genetics Society of America.

1951 First public presentation of transposable elements, at a Cold Spring Harbor symposium, is greeted with "puzzlement, even hostility."

1967 Becomes Distinguished Service Member of the Carnegie Institution at Cold Spring Harbor; receives Kimber Genetics Award.

1970 Receives National Medal of Science from President Richard M. Nixon.

1973 Dedication of the McClintock Laboratory at Cold Spring Harbor.

1981 Within two months, becomes first recipient of a MacArthur Foundation grant and receives Albert Lasker Award for Basic Medical Research and Wolf Prize in Medicine.

1983 Receives the Nobel Prize in Physiology or Medicine on October 10.

1992 Death of Barbara McClintock at Cold Spring Harbor on September 2.

Bibliography

"Barbara McClintock, Won Nobel Prize for 'Jumping Genes' Discovery." *The Boston Globe*, September 4, 1992: 59. Available online at *www.boston.com/globe/search/stories/nobel/1992/1992k.html*.

Buckler Lab of Plant Genomics and Diversity (North Carolina State University, Department of Genetics). *www.maizegenetics.net*.

Comfort, Nathaniel C. *The Tangled Field: Barbara McClintock's Search for the Patterns of Genetic Control*. Boston: Harvard University Press, 2001.

Fedoroff, Nina, and David Botstein, eds. *The Dynamic Genome: Barbara McClintock's Ideas in the Century of Genetics*. Cold Spring Harbor: Cold Spring Harbor Laboratory Press, 1992.

Johannsen, Wilhelm Ludwig. "The Genotype Conception of Heredity." *American Naturalist* XLV: 132 (1911): 129–159.

Keller, Evelyn Fox. *A Feeling for the Organism*. New York: W.H. Freeman, 1983.

NLM (National Library of Medicine), profile of Barbara McClintock. Available online at *profiles.nlm.nih.gov/LL/Views/Exhibit/narrative/nobel.html*.

The letters quoted in this text can be found online as follows:

Fincham: *profiles.nlm.nih.gov/LL/B/B/G/C/_/llbbgc.pdf*

Goddard and Stern: *profiles.nlm.nih.gov/LL/B/B/M/P/_/llbbmp.pdf*

Nelson: *profiles.nlm.nih.gov/LL/B/B/F/X/_/llbbfx.pdf*

Nobel Assembly. Several citations and diagrams used in the text, taken from a press release and from Dr. McClintock's Nobel autobiography, were produced by the Nobel Assembly (at the Karolinska Institute in Sweden) at the time of the award. These are now available online at the Nobel E-Museum (*www.nobel.se*).

Shugurensky, Daniel. *History of Education: Selected Moments of the 20th Century*. Department of Adult Education and Counselling Psychology, The Ontario Institute for Studies in Education of the University of Toronto (OISE/UT). Available online at *fcis.oise.utoronto.ca/~daniel_schugurensky/assignment1/1919pea.html*.

Further Reading

Comfort, Nathaniel C. "Barbara McClintock's Long Postdoc Years." *Science* 295 (January 18, 2002): 440.

Dash, Joan. *The Triumph of Discovery: Women Scientists Who Won the Nobel Prize.* Englewood Cliffs, NJ: Julian Messner, 1991.

Hall, Mary Harrington. "The Nobel Genius." *San Diego Magazine,* August 1964.

Hawking, Stephen, ed. *On the Shoulders of Giants: The Great Works of Physics and Astronomy.* Philadelphia: Running Press, 2002.

McGrayne, Sharon Bertsch. *Nobel Prize Women in Science.* New York: Birch Lane Press, 1995.

———. *Women in Science: Their Lives, Struggles and Momentous Discoveries.* Secaucus, NJ: Carol Publishing Group, 1998.

———. "McClintock and Marriage." *Science* 2002 April 5; 296:47 (in Letters).

Peterson, Thomas. "A Celebration of the Life of Dr. Barbara McClintock." *Probe* (newsletter of the USDA Plant Genome Research Program) 3:1/2 (January–June 1993).

Yount, Lisa. *Twentieth-Century Women Scientists.* New York: Facts on File, 1996.

Websites

American Philosophical Society: The Barbara McClintock Papers
www.amphilsoc.org/library/browser/m/mcclintock.htm

Cold Spring Harbor Laboratory
www.cshl.org

U.S. Department of Energy: Genomics and Its Impact on
Genetics and Society: A Primer
www.ornl.gov/hgmis/publicat/primer2001/index.html

Genetics in Context
www.esp.org/timeline/

MendelWeb
www.mendelweb.org

U.S. Department of Energy: Office of Science: Office of Biological and
Environmental Research: The Human Genome Project
www.er.doe.gov/production/ober/hug_top.html

National Library of Medicine: "The Barbara McClintock Papers"
profiles.nlm.nih.gov/LL/Views/Exhibit/

Nobel E-Museum
www.nobel.se/

U.S. Department of Agriculture: Agricultural Research Service: National
Agricultural Library: Plant Genome Data & Information Center
www.nal.usda.gov/pgdic/

The MacArthur Foundation
www.macfound.org/

A Primer on Molecular Genetics
www.gdb.org/Dan/DOE/intro.html

Biology Online
www.biology-online.org

The Association for Women in Science
1200 New York Ave., Suite 650 NW
Washington, DC USA 20005
202.326.8940
www.awis.org

Index

Activator, 82-83
American Philosophical Society of Philadelphia, 107
Andrew White Professor-at-Large (Cornell University), 94-95

Beadle, George W., 50, 54-55, 76-78, 89-90, 93, 98, 99
Berlin, Germany, and Kaiser Wilhelm Institute, 61
Blumenschein, Almeiro, 93
Botanical Institute (Freiburg), 61-62
Breakage-fusion-bridge cycle, 65-68, 109
Brink, R.A., 84
Brooklyn, New York, 13-14, 23-29
Brooklyn Public Library, 29
Brussels sprouts, 88

Cabbage, 88
California Institute of Technology (Cal Tech), 51, 57, 59, 64, 76, 87
Cancer, 71, 109
Carnegie, Andrew, 90
Carnegie Institution of Washington at Cold Spring Harbor, McClintock at, 72-76, 79-80, 82-87, 90, 91, 93, 100, 104-106, 107
Cauliflower, 88
Centromeres, 65-67
Chromosomal Constitution of Races of Maize (McClintock, Blumenschein, and Kato), 93
Chromosomes, 36-37, 39-41
   and breakage-fusion-bridge cycle, 65-68
   broken, 56-59, 65-71
   and centromeres, 65-67
   of corn, 45-49, 50, 52, 54, 56
   and crossing over, 40-41, 44, 54
   and genes, 40-41

and meiosis, 37, 54, 65-67, 77
   morphology of, 48
   and mutations, 56-59, 65-71
   ring, 56-59, 65
   and sex-linked characteristics, 37, 39-40
   and telomeres, 68-71
   See also Maize, McClintock's work on genetics of
Cloning, 109
Cold Spring Harbor. *See* Carnegie Institution of Washington
Columbia, McClintock's research on corn in, 91-92
Controlling elements. *See* Mobile genetic elements
Corn. *See* Maize, McClintock's work on genetics of
Cornell University
   and McClintock as research associate, 50-52, 54-55, 62
   and McClintock as student, 14, 30-32, 41, 43-49
   and McClintock as Andrew White Professor-at-Large, 94-95
   and McClintock as distinguished alumna, 107
Corn on the cob, origins of, 87-90, 91-93
   See also Maize, McClintock's work on genetics of
"Correlation of Cytological and Genetically Crossing-Over in *Zea Mays*, A," 54
Creighton, Harriet, 52, 54, 60
Crick, Francis H.C., 96-97
Crossing over, 40-41, 44, 54
Curie, Marie, 15

Deletion, 57
Demerec, Milislav, 72-73, 76, 85

Deoxyribonucleic acid (DNA), 96-97, 107

De Vries, Hugo, 34

Dissociator, 82

Distinguished Service Award (Carnegie Institution of Washington), 93, 94

Doctor of Science degree, 49 honorary, 95-96

Dominant genes, 33-34, 36

Dynamic Genome: Barbara McClintock's Ideas in the Century of Genetics, The (Fedoroff), 106

Dynamite, and Nobel, 97, 99

Ebert, James D., 100

Emerson, Rollins, 45, 59, 64

Ephrussi, Boris, 76-77

Erasmus High School, 27-29

Fedoroff, Nina, 106

Freiberg, Germany, Botanical Institute in, 61-62

Fruit flies *(Drosophila melanogaster)*, 37, 39-40, 45, 51, 56, 60, 76-77, 83, 85

Gene mapping, 108-109

Genes
dominant, 32-34, 36
as real objects on chromosomes, 40
recessive, 32-34, 36
as term, 40
and X-rays, 56-59
See also Chromosomes; Genetics

Genetic engineering, 109

Genetics
and crossing over, 40-41, 44, 54
and fruit flies, 37, 39-40, 45, 51, 56, 60, 76-77, 83, 85
and Mendel, 32-34, 36, 37, 85

and Morgan, 37, 39-41
and one gene-one enzyme concept, 76-78, 99
and sex-linked characteristics, 37, 39-40
See also Chromosomes; Genes; Maize, McClintock's work on genetics of; Mutations

Genetics, 69, 76

Genetics Society of America, McClintock as president of, 79

Geriatrics, and telomeres, 71

Germany, McClintock's research in, 59-62

Goldschmidt, Richard B., 61-62

Great Depression, 56, 64

Guggenheim Research Fellowship, 59

Handy, Benjamin (grandfather), 18, 19, 20

Handy, Hatsel (great-grandfather), 18

Henking, Hermann, 36

Heredity. *See* Genetics

Hitler, Adolf, 60-61, 62

Hodgkin, Dorothy Crowfoot, 15

Hofmeister, Wilhelm, 36

Human Genome Project, 108-109

Hutchinson, C.B., 43-44

Immunology, 109

Intuitive leaps, 34

Israel, and Wolf Foundation grant, 97

Johannsen, Wilhelm, 40

Jumping genes, 83

Kaiser Wilhelm Institute, 61

Kale plants, 87-88

Kato, Angle, 93

Index

Keller, Evelyn Fox, 22-23
Kimber Genetics Award (National
    Academy of Sciences), 94
Kissinger, Henry, 107

Lasker Award for Basic Medical
    Research, 97
Lederberg, Joshua, 85, 99

MacArthur Foundation grant, 97
McClintock, Barbara
    and awards and honors, 12, 14,
        15, 17, 56, 59, 79, 93, 94-96,
        97, 99-100
    birth of, 20
    childhood of, 13-14, 20, 21, 22-26
    death of, 12, 106-107
    and doctorate, 49
    and early interest in science,
        22-23
    and early jobs, 29
    education of, 24-25, 27-29, 30-32,
        41, 43-49
    and Eleanor as birth name, 20
    family of, 13-14, 17, 18-24, 25,
        28-29
    and fellowships, 56, 59
    and friends and colleagues, 51-53,
        56-57, 59, 60-61, 64-65, 72-
        76, 91, 93, 99, 104-106
    and funding, 56, 59, 64
    and Genetics Society of America,
        79
    in Germany, 59-62
    as "Katharine Hepburn of
        Science," 100, 103
    legacy of, 12, 14, 104-109
    and master's degree, 49
    and name change from Eleanor
        to Barbara, 21
    and National Academy of
        Sciences, 79

and Nobel Prize in Physiology
    or Medicine, 12, 14, 15, 17,
    99-100, 103
personality of, 13-15, 17, 21, 22,
    25-26, 41, 43, 65, 71, 84, 100,
    103, 106-107
as professor at University of
    Missouri, 64-71
and publications, 49, 54, 69, 76,
    83, 93
and repository for papers, 107
and reproductive cycle of fungus,
    77-79
as research associate at Cornell,
    50-52, 54-55, 62
and Rockefeller Foundation
    grant, 64
and skill with microscope, 44-45,
    48-49, 57, 77-78
as spinster and career woman,
    62-63
and travels, 12, 14, 56, 59-62,
    77-79, 86-87, 91-93
See also Maize, McClintock's
    work on genetics of
McClintock, Malcolm Rider
    ("Tom") (brother), 20-21
McClintock, Marjorie (sister), 20,
    27-28
McClintock, Mignon (sister), 20
McClintock, Sara Handy (mother),
    13-14, 18-20, 21, 22, 24, 25-26,
    28-29, 30
McClintock, Thomas Henry
    (father), 13-14, 20, 21-22, 23, 24,
    25, 29, 30
Maize, McClintock's work on
    genetics of, 43-49, 50-52
    and breakage-fusion-bridge
        cycle, 65-68, 109
    and broken chromosomes, 56-59,
        65-71

at Carnegie Institute of
Washington, 72-76, 79-80,
82-87, 90, 91, 93, 100, 104-
106, 107
at Cornell, 45-49, 50-52, 54-55, 62
and crossing over, 54
and genes contained in chromosomes, 48-49
and identifying chromosomes,
48, 54
and mobile genetic elements, 12,
14, 15, 17, 79-80, 82-87
and origins of corn on the cob,
90, 91-93
and planting maize, 52, 54
and ring chromosomes, 56-59
and telomeres, 68-71
at University of Michigan, 64-71
Mangelsdorf, Paul Christof, 88-90,
91
*Mechanism of Mendelian Heredity,
The* (Morgan), 41
Meiosis, 37, 54, 65-67, 77
Mendel, Gregor, 32-34, 36, 37, 85
Mexico, and origin of corn, 87-90
Mobile genetic elements, 12, 14, 15,
17, 79-80, 82-87
Molecular biology, 96
Montgomery, Thomas H., 36
Morgan, Thomas Hunt, 37, 39-41,
56, 59, 64, 76
Muller, Herman, 56
Mutations
and breakage-fusion-bridge
cycle, 65-68
and mobile genetic elements, 12,
14, 15, 17, 79-80, 82-87
and X-rays, 56-59, 68

National Academy of Sciences,
79, 94

National Library of Medicine, 107
National Medal of Science, 96
National Research Council, 56
Nazis, 60-61, 62
Nelson, Oliver, 96, 106
Neurospora mold, 76-79
Nixon, Richard M., 96
Nobel, Alfred, 97, 99
Nobel Prize, 97, 99
to Beadle, 78, 99
to Crick, 68, 97
to Lederberg, 85, 99
to McClintock, 12, 14, 15, 17,
99-100, 103
to Morgan, 41
to Muller, 56, 58
to Tatum, 78, 99
to Watson, 68, 97, 99
to Wilkins, 97

One gene-one enzyme concept,
78, 99

Peru, McClintock's research on
corn in, 91
Peterson, P.A., 84
*Proceedings of the National Academy
of Sciences,* 54, 83

Radiation sickness, 109
Recessive genes, 33-34, 36
Rhoades, Marcus M., 50-51, 54-55,
72-73, 74, 99
Ring chromosomes, 56-59, 65
Roaring Twenties, 43
Rockefeller, John Davidson, 90
Rockefeller Foundation, 64, 65,
90-93
Rückert, Johannes, 36
Sex-linked characteristics, 37, 39-41
Shapiro, James, 106

Index

Sex-linked characteristics, 37, 39-41
Shapiro, James, 106
Sharp, Lester, 52
Shoot-bagging, 47
South America, McClintock's
    research on corn in, 91-93
"Stability of Broken Chromosomes
    in *Zea Mays,* The," 68-69
Stadler, Lewis, 56-57, 59, 64-65, 71
Stanford University, 77-79
Stern, Curt, 60-61, 62, 93
Sundaresan, V., 107
Sutton, Walter Stanborough, 36-37

Tatum, Edward, 77-78
Telomeres, 68-71
Teosinte hypothesis, 88
Transposable elements. *See* Mobile
    genetic elements

Tripartite hypothesis, 88-89
Tripsacum, 88-89

University of Missouri, 64-71

Von Tschermak, E., 34

Waldeyer, Heinreich, 36
Watson, James D., 96-97, 99, 100,
    107
Weaver, Warren, 96
Wilkins, Maurice Hugh Frederick,
    97
Witkin, Evelyn, 106-107
Wolf Foundation, 97
World War I, 29, 41
World War II, 74, 75

X-rays, and mutations, 56-59, 68

Picture Credits

Contributors

J. HEATHER CULLEN has spent her career working in scientific and medical publishing. She holds a B.S. degree in nutrition from Cornell University and an M.S. degree in technical and scientific communication from Drexel University. She is a member of Soroptimist International of the Americas, a professional women's organization that seeks to improve the status of women throughout the world. She lives and writes in historic Philadelphia with her golden retriever, Angie.

JILL SIDEMAN, PH.D. serves as vice president of CH2M HILL, an international environmental-consulting firm based in San Francisco. She was among the few women to study physical chemistry and quantum mechanics in the late 1960s and conducted over seven years of post-doctoral research in high-energy physics and molecular biology. In 1974, she co-founded a woman-owned environmental-consulting firm that became a major force in environmental-impact analysis, wetlands and coastal zone management, and energy conservation. She went on to become Director of Environmental Planning and Senior Client Service Manager at CH2M HILL. An active advocate of women in the sciences, she was elected in 2001 as president of the Association for Women in Science, a national organization "dedicated to achieving equity and full participation for women in science, mathematics, engineering and technology."